New Trends in Banking Management

Contributions to Management Science

Harald Dyckhoff/Ute Finke
Cutting and Packing in Production and Distribution
1992. ISBN 3-7908-0630-7

Richard Flavell (Ed.)
Modelling Reality and Personal Modelling
1993. ISBN 3-7908-0682-X

Michael Hofmann/Monika List (Eds.)
Psychoanalysis and Management
1994. ISBN 3-7908-0795-8

Rita L. D'Ecclesia/Stavros A. Zenios (Eds.)
Operations Research Models in Quantitative Finance
1994. ISBN 3-7908-0803-2

Mario S. Catalani/Giuseppe F. Clerico
Decision Making Structures
1996. ISBN 3-7908-0895-4

M. Bertocchi/E. Cavalli/S. Komlósi (Eds.)
Modelling Techniques for Financial Markets and Bank Management
1996. ISBN 3-7908-0928-4

Holger Herbst
Business Rule-Oriented Conceptual Modeling
1997. ISBN 3-7908-1004-5

Constantin Zopounidis (Ed.)
New Operational Approaches for Financial Modelling
1997. ISBN 3-7908-1043-6

Klaus Zwerina
Discrete Choice Experiments in Marketing
1997. ISBN 3-7908-1045-2

Giovanni Marseguerra
Corporate Financial Decisions and Market Value
1998. ISBN 3-7908-1047-9

WHU Koblenz – Otto Beisheim Graduate School of Management (Ed.)
Structure and Dynamics of the German Mittelstand
1999. ISBN 3-7908-1165-3

Armin Scholl
Balancing and Sequencing of Assembly Lines
1999. ISBN 3-7908-1180-7

Elio Canestrell (Ed.)
Current Topics in Quantitative Finance
1999. ISBN 3-7908-1231-5

W. Bühler/H. Hax/R. Schmidt (Eds.)
Empirical Research on the German Capital Market
1999. ISBN 3-7908-1193-9

M. Bonilla/T. Casasus/R. Sala (Eds.)
Financial Modelling
2000. ISBN 3-7908-2282-X

S. Sulzmaier
Consumer-Oriented Business Design
2001. ISBN 3-7908-1366-4

Constantin Zopounidis
Editor

New Trends in Banking Management

With 43 Figures
and 42 Tables

Physica-Verlag
A Springer-Verlag Company

Series Editors
Werner A. Müller
Martina Bihn

Author
Professor Constantin Zopounidis
Technical University of Crete
Department of Production Engineering and Management
University Campus
73100 Chania
Greece
kostas@ergasya.tuc.gr
kostas@cha.forthnet.gr

ISSN 1431-1941
ISBN 3-7908-1488-1 Physica-Verlag Heidelberg New York

Die Deutsche Bibliothek – CIP-Einheitsaufnahme
New trends in banking management: with 42 tables/Constantin Zopounidis, ed. – Heidelberg; New York: Physica-Verl., 2002
(Contributions to management science)
ISBN 3-7908-1488-1

Physica-Verlag Berlin Heidelberg New York
a member of BertelsmannSpringer Science+Business Media GmbH

http://www.springer.de

Softcover design: Erich Kirchner, Heidelberg

SPIN 10874697 88/2202-5 4 3 2 1 0 – Printed on acid-free paper

Dedicated to my son Dimitris

Words, by the absence of acts, seem to be futile and unavailing

a) DIMOSTHENIS 1. Olinthiakos B12
2. Epistoli Filippou 23

b) AISCHILOS (Prom. Desmotis 1080)

Preface

During the last decades the globalization, the intensified competition and the rapid changes in the socio-economic and technological environment had a major impact on the global economic, financial and business environments. Within this environment, it is clear that banking institutions worldwide face new challenges and increasing risks, as well as increasing business potentials. The recent experience shows that achieving a sustainable development of the banking system is not only of interest to the banking institutions themselves, but it is also directly related to the development of the whole business and economic environment, both at regional and international level.

The variety of new banking products that is constantly being developed to accommodate the increased customer needs (firms, organizations, individuals, etc.) provides a clear indication of the changes that the banking industry has undergone during the last two decades. The establishment of new products requires the implementation of innovative processes and instruments for their efficient management. The implementation of such processes and instruments is closely related to a variety of disciplines, advanced quantitative analysis for risk management, information technology, quality management, etc. The implementation of these approaches in banking management is in accordance with the finding that empirical procedures are no longer adequate to address the increasing complexity of the banking industry.

The above issues were the main motivation for organizing the 10th Special Conference of the International Federation of Operational Research Societies (IFORS) during April 1-3, 2001 in Athens, Greece, under the theme "New Trends in Banking Management". The objective of the conference was to bring together researchers and practitioners from all around the world to present and discuss the emerging topics of interest to banking management. Special focus was given to the advances made on the methodological tools that can be used to improve the operation, effectiveness and performance of the banking sector.

The present edited volume includes some of the papers that were prepared during the above conference. After a rigorous reviewing process, eighteen papers were finally accepted for publication in the special issue. These papers cover a wide range of banking problems, including bank performance, efficiency and viability, credit risk assessment, bank marketing, quality management, bank mergers and acquisitions, logistics in banking, portfolio management as well as mutual funds management.

The first part of the book includes three papers on banking quality, efficiency/profitability and customer satisfaction analysis. The first paper by Spathis, Petridou and Glaveli presents the results of an empirical study using the banking service quality model to investigate the service quality of Greek banks. The paper of Anastasiou, Mitropoulos and Sissouras employs the data envelopment analysis to assess the efficiency of 39 bank branches of the

Commercial Bank of Greece and to specify appropriate efficiency targets. The third paper of this section by Bouranta, Kouremenos and Siskos presents the results of a pilot customer satisfaction survey on the quality of services provided by bank ATMs and tellers in Greece. The results of the survey are analyzed using a multicriteria decision-aid methodology. The multicriteria decision-aid approach is also employed in the paper of Stathas, Kosmidou, Doumpos and Zopounidis. The authors employ the PROMETHEE II method to evaluate the financial performance of banks in Greece.

The second part of the book involves business failure prediction and credit risk assessment. The first paper by Daubie and Meskens reviews the existing literature on business failure prediction and discusses some important issues on the problem, such as the causes of business failure, its symptoms and the possible remedies. The second paper by Michalopoulos, Dounias, Hatas and Zopounidis addresses the credit risk assessment problem. The authors propose a methodology that combines fuzzy logic (a fuzzy c-means clustering algorithm) with neural networks. This methodology is applied on a sample of 130 credit applications to develop a credit scoring model.

The third part of the book consists of three papers on banking organization and customer relationship management. The first paper of this part, by Troege, provides a short and synthetic review of the literature on banking relations, underlining the implications for bank management and regulatory policy. The second paper by Karellas and Stamoulis discusses the organization model of a bank's branch network and proposes a model to decide on the focus and allocation of banking units spread within a region. The last paper of this third part by Vlachopoulou, Manthou and Ioannidis, develops a modeling framework for customer relationship management (CRM) in the banking sector.

The fourth part of the book includes three papers on the reforms in the banking sector including mergers and acquisitions. The first paper by Gupta and Chevalier analyzes the takeover battle between three French banks (BNP, Société Générale and Paribas) providing a critical analysis of this battle and comments on the expected synergies. The second paper by Singh and Dhillon presents the Indian experience on reforms and the restructuring of the banking sector focusing on strategy and performance issues. The third paper by Staikouras and Wood investigates the sensitivity of banks' income sources (net interest margin) to changes in nominal interest rates. For this purpose cross-country time-series data are employed from European Union countries, involving a large sample of more than 2,000 banks.

Portfolio management and risk assessment are the focus in the fifth part of the book, which consists of four papers. The first paper by Luciano and Marena provides value at risk bounds for portfolios of correlated financial assets using data involving 6 international stock market indices. Value at risk is also the focus in the second paper by Barucci and Renò. The authors develop a model to forecast daily volatility of exchange rates in order to be able to calculate reliable value at risk estimates. A comparison with traditional models is also presented. The third paper by Michalopoulos, Zopounidis and Stavroulakis uses data from the Athens Derivatives Exchange to examine the timing, profitability and risk of index arbitrage transactions through futures contacts. The last paper of this part by Pendaraki and Zopounidis provides a comprehensive review of the existing literature on mutual funds' performance evaluation focusing on a critical analysis of traditional performance measures.

The last part of the book includes two papers on risk logistics in banking and investment analysis. The first paper by Tarantillis and Kiranoudis presents an adaptive memory-based heuristic algorithm that minimizes the risk of robbery of a fleet of armored vehicles, departing and terminating at the central offices of a bank serving a set of bank delivery offices. The second paper of this part and the final paper of the book by Sambracos presents a real-world case study regarding the risk analysis of an investment project involving the development of a new suburban coastal transport system in Athens, Greece.

Of course, one should acknowledge that the field of banking is a very broad one, such that it would be impossible to address it in a fully comprehensive way within the limits of a single edited volume. Nevertheless, the range of banking problems that are addressed in the papers of this book, the variety of methodological tools that are employed, and the combination of the theoretical developments with their applications in practice, make this book a notable reference for academic researchers and practitioners who are involved with the modern banking management.

Sincere thanks must be expressed to all the authors whose contributions have been essential in creating this book. I also owe a great debt to those who worked long and hard to review all the submitted papers and contributed to the achievement of a high standard of this edited volume. Special thanks should also be given to Dr. Michael Doumpos for his valuable assistance in my contacts with the authors and his help in the material collection and management.

Constantin Zopounidis
Chania, Greece
February 2002

Contents

Banking Quality, Efficiency, Profitability and Customer Satisfaction Analysis

Business Failure and Credit Risk Assessment

Banking Organization, Banking Marketing and Customer Relationship Management

Bank Mergers, Acquisition and Concentration

Portfolio Management and Risk Assessment

Risk Logistics in Banking Management

Banking Quality, Efficiency, Profitability and Customer Satisfaction Analysis

An Empirical Study of Service Quality Perspectives in Public and Private Banks

Ch. Spathis[1], E. Petridou[1], N. Glaveli[2]

[1] Aristotle's University of Thessaloniki, Department of Economics, Division of Business Administration, 54006 Thessaloniki, Greece

[2] University of Macedonia, Department of Business and Administration, 54006 Thessaloniki, Greece

Summary:
During the past decade the banking sector has undergone drastic changes, resulting in a high competitive market place. The resulting margin squeeze and disintermediation have emphasised the need for quality. A proper understanding of the determinants of quality can be seen to have an increasing interest for banks in the competitive environment. The present empirical study undertaken in Greek banks investigated their service quality based on customers' perceptions as measured with the Banking Service Quality (BSQ) model. Findings suggest that the perception of the profile of services received was higher in the private than in the public sector in all factors.

Keywords: Banking Quality, Private – Public Banks, Quality Management, Factor Analysis, Greece

1 Introduction

During the last decade, the Greek banking system undergone drastic changes such as: (a) deregulation, (b) creation of extensive product/service portfolios (c) major changes in the ownership status (d) heavy use of modern technology and (e) globalisation of the Greek banks activities (Gortsos, 1998). With the establishment of Euro, competition will be further intensified in Pan - European level. The effects on the Greek banking system, from the establishment of Euro are expected to have a high adjustment cost that will affect its profitability (Kosmidou and Spathis, 2000). Due to heightened competition, bank service quality rises as an important factor that will affect the relevant market shares and profitability in the banking sector (Anderson *et al.*, 1994; Hallowell, 1996; Caruana and Pitt, 1997, Spathis *et al.*, 2001).

Quality is sought by all organisations, especially in the service sector. This is of main interest, particularly, in the banking sector. Greek banks although they declare their focus on quality and customer-oriented strategies have not used, besides some sporadically surveys, a publicly recognised standard scale to measure the perceived quality of their services. The resulting margin squeeze has emphasised the need for Greek banks to sell more products (cross-selling), adopt new dynamic marketing strategy, develop new innovative products and place higher emphasis on both the tangible and intangible aspects of their services (Chetham, 1994; Kloutsiniotis, 1995; Noulas, 1999; Petridou and Glaveli, 2000). Multipurpose banking, customer oriented distribution training for reskilling employees and heavy use of technology, are the main strategies that were adopted to better satisfy the changing customer needs by improving their perceived quality (Moschos and Fragetis, 1997).

Empirical evidence has shown that service quality can be seen as one of the main determinants of customer satisfaction, which in turn *may* influences purchase intentions (Cronin and Taylor, 1992; Vavra *et al., 1995, 1997;* Pruden *et al., 1996;* Spreng and Mackoy, 1996; De Ruyter, 1997; Bloemer *et al.*, 1998). Practitioners and academics alike are keen on accurately measuring service quality in order to better understand its essential antecedents and consequences, and, ultimately establish methods for improving quality to achieve competitive advantage and build customer relationships (Zahorick and Rust, 1992; Palmer and Cole, 1995). Zopounidis *et al.,* (1995; 1997) evaluated the efficiency of the banking branches of the Bank of Greece with the aid of the multicriteria approach UTADIS.

The objective of this study is the investigation of the Greek Public and Private Banks quality of service based on their customers' perceptions. The Bank Service Quality (BSQ) model developed by Bahia and Nantel (2000) has been used for determining and ranking the service quality factors. Subsequently, a comparison has been made between the factors influencing the public and private bank service quality.

2 Literature Review

Quality has been defined as superiority or excellence, or, as the consumer's overall impression of the relative inferiority/superiority of the organisation and its services (Zeithaml *et al.,* 1990; Bitner and Hubbert, 1994; Taylor and Baker, 1994). Perceived service quality is a form of attitude, a consumer judgement of the overall excellence or superiority of the service. Perceived service quality results from a comparison of consumers' expectations with their perceptions of service delivered by the supplier (Kagnis and Voukelatos, 1997). Thus, in the banking sector, perceived service quality results from the difference between customers' perceptions for the service offered and their expectations vis-a-vis the bank offering the expected service.

If perceptions are not meeting or exceeding expectations, a gap is created which should be addressed through strategies targeting to affect either expectations or perceptions or even both (Parasuraman *et al.*, 1985; Zeithaml *et al.,* 1990). Expectations are formed from a variety of sources that could be grouped in two main categories. First, sources associated with consumers' characteristics - such as personal needs and wishes (Edvardsson *et al.,* 1994) personal philosophy about a particular service, past experience and beliefs of other person's. Second, sources associated with the organisation such as promises by staff, advertising, implicit service promises (such as price, image and reputation, in service encounters verbal or non verbal behaviour) evidence of service (employees, process and physical evidence) and distribution channels (Zeithaml and Britner, 1996).

Various models have been developed for measuring perceived service quality. The SERVQUAL model (Parasuraman *et al.,* 1985; 1988) and the Technical / Functional Service Quality framework (Gronoos, 1983; 1990) are the most widely accepted and tested ones. Bahia and Nantel (2000), developed the Banking Service Quality- BSQ model, a model developed especially for the banking sector.

Parasuraman *et al.*, (1985) based on the expectations-performance gap analysed above, developed a ten-dimension mode, the SERVQUAL model. Later in 1988, the original model was replaced by a five-dimension construct of perceived service quality. The model includes the following factors: (a) tangibles, which are the physical evidence of the service; (b) reliability, involving consistency of performance and dependability; (c) responsiveness, concerning the willingness and readiness of employees to provide service; (d) assurance, knowledge and courtesy of employees and ability to inspire trust and confidence; and (e) empathy, pertaining to carry individualised attention that a firm provides its customers. Generally the SERVQUAL approach has not been without its critics. Studies support that the five-dimensions of the SERVQUAL approach may not be universal across all services (Brensinger and Lambert, 1990; Babakus and Boller, 1992). Moreover, these dimension are not generic. Carman (1990) for instance, found that it is often necessary to incorporate additional items for some service categories. SERVQUAL has been deemed "to be applicable to retail and business services and relevant for banking services" (Zeithaml and Bitner, 1996). Further criticism is addressed because it places emphasis mainly on the service / product dimension and tends to neglect other significant dimensions of the 7Ps' marketing mix (Gilmore and Carson, 1992).

Another well-accepted model of service quality is the Technical / Functional Quality perspective (Arora and Stoner, 1996). As originally conceptualised by Gronoos (1983) the Technical / Functional framework, distinguishes between the technical quality (what is provided) and the functional quality (how it is provided). Meeting customers requirement (i.e is the service carried out in a way that meets customers requirements?) as well as conformity to measurable standards (i.e. does the service meet the promised standards?) were the basis of Gronoos (1983) identification of Technical / Functional quality perspectives. This model fails to address an important element, the extent to which, over time, the service can be determined in conformity with the "best" possible service that could have been provided (Davies e*t al.*, 1999). It means that a service has to meet the twin tests of performance (if it has been delivered in a way that was acceptable, and if it met the standards expected or had the desired results) whilst a different service might had resulted in superior outcomes for the customer. Although the Technical / Functional Quality Model has not been utilised or tested to the extent of the SERVQUAL model, it has received some research / empirical attention in the last years (Higgins and Ferguson, 1991; Richard and Allaway, 1993; Baker and Lamp, 1993).

The BSQ model (Bahia and Nantel, 2000) is an extension of Parasuraman's *et al.,* (1985) model. Bahia and Nantel, conceptualised this model and tested it in a study performed in Canada, in order to develop a reliable and valid scale of the perceived service quality of banks services. The population from which their sample of 115 respondents was selected was formed by 18 to 60 years – old French speaking customers of the National Bank of Canada. Bahia and Nantel referred to Parasuraman's original 10 dimensions, since in its five dimensions the SERVQUAL model has not been proven stable across replications. By reconsidering the ten dimensions and by adding items that cover all the facets of the marketing mix (Boom and Britner, 1981) they managed to avoid the problem of some dimension being insufficiently generic and contributed to construct and nomological validities for courtesy and access. They ended up with a proposed scale (the BSQ scale) that comprises 31 items, which span six dimensions: (a) effectiveness and assurance, (b) access, (c) price, (d) tangibles, (e) service portfolio, and (f) reliability. Comparing BSQ with SERVQUAL, Bahia and Nantel (2000) argue that the first advantage of BSQ for the banks is related to its content validity. In fact BSQ is exclusively dealing with the perceived service quality within the special context of banks. Moreover, the item's inclusion in the different dimensions of the construct is much more consistent. The BSQ model was used for measuring the perceived service quality of banks. As stated above, the BSQ model was chosen as more appropriate for the banking sector than the SERVQUAL and the Technical / Functional Quality model.

The underlying reason for seeking service quality is a belief that service quality can lead to increased customer satisfaction and thus, contribute significantly to performance (Berry *et al.,* 1985; Barnes and Cumby, 1995). In a meta analysis carried out by Capon *et al.,* (1990) they identified 20 studies that find positive relationship between quality and business performance. Service quality has been linked to performance both through activities for customer retention (Reichheld and Sasser, 1990) and through management expectations and service reliability - two key components of service quality - (Edvardsson and Gustavsson, 1990; Berry and Parasuraman, 1991). In a large-scale examination of Swedish firms, Anderson *et al.,* (1994), confirm that firms that actually achieve high customer satisfaction (due to high quality perceptions) also enjoy superior economic returns. They found that an annual one-point increase in customer satisfaction results in a net present value that represents a cumulatively increase of 11.5 per cent over five years for a typical firm in Sweden. Rust *et al.,* (1994), who document the marketing literature relating customer satisfaction and service quality to performance, conclude that "a link between quality and financial returns exists and the challenge is to provide operational measures for measuring the link".

In the Greek banking sector, despite the emphasis given on quality, there is a gap of research on the perceived service quality of the Greek bank customers. Despite the survey conducted by Kagnis and Voukelatos (1997) there is no other similar survey (known to authors). They measured, with the use of the SERVQUAL model, and compared customer expectations and perceptions on bank (private and public) service quality. The research was located only in the area of Athens and the sample consisted of 163 respondents. The results have showed that quality expectations and evaluation of services received were marginally higher in the private sector than in the public sector in most dimensions measured. The dimensions of the SERVQUAL model were rank as follows: Tangibles, Reliability, Responsiveness, Assurance and Empathy.

Another relevant Greek study, which is worth mentioning, is the one by Mihelis *et al.,* (2001). They conducted an original satisfaction survey in the private bank sector. With the use of multicriteria analysis and preference desegregation modelling they determined the critical service dimensions and segmented the market to customer clusters with distinctive preferences and expectations. The main satisfaction criteria used were: personnel, products, service and access and image. 'Access" and "Personnel" appear with the highest satisfaction indices.

3 Methodology

3.1 The Sample

The main objectives of our survey are to detect the bank customers' perception of service quality, in private and state-owned banks in Greece and examine the quality environment in which profitability flourishes. The sample consisted of 1260 private individual customers of either public or private banks in the area of Greece (it covers the big urban centers and the rural areas as well). As the purpose of the research was not to measure quality in relation to any particular bank, respondents were approached randomly and were grouped by a post-stratification method (Curwin and Slater, 1991). The sample characteristics are summarised in Table 1. The sample was conducted by personal interviews at the branches. The customers were asked to express their opinion of the bank that they had just completed a transaction with. Initially "intercept interviews" were conducted. Data collection was over a period of 30 days, at different times and locations with a view to sampling a wide range of customers.

A pre-tested prompt was used to ensure consistency in the enquiry; the questions were close-end. After a pilot study with a convenience sample of 60 respondents, the instrument was slightly corrected. We corrected mainly the phrasing of the questions, for improving respondents understanding. In measuring perceptions of quality received, respondents were asked to state the degree to which their bank performed the feature described. The questionnaire consisted of two parts: the first part contained statements about service quality perceptions taken from the original BSQ model questionnaire used in Bahia's and Nantels' (2000) survey. The second part aimed at building a demographic profile of customer use of banks, status, city, gender, age and income.

3.2 Method

The statistical analysis methods employed were: (a) t-test for significance of the means of the value by variable, (b) factor analysis for the evaluation of the scale of measurement for quality perspectives. Factor analysis was used for two main reasons: First, for synthesising the characteristics into a smaller number of principal factors and second because it was used by Bahia and Nantel (2000) in their original survey, thus comparison between the results was applicable.

Factor analysis includes a group of multivariate statistical methods, the principal aim of which is the analysis of the properties of the data (variables) and their presentation as a smaller number of composite factors with a minimal loss of information (Hair *et al.*, 1992; Zimkund, 1994). In the mathematical model of the factor analysis, each variable expresses a linear combination of factors. In a general form the model for the i standardised variable X_i could be written:

$$X_i = A_{i1} F_1 + A_{i2} F_2 + \dots\dots\dots + A_{ik} F_k + U_i$$

where F_j (j =1, 2,.......,k) are the common factors, where all the variables are expressed as correlations with these, U is the unique factor representing the part of the variable X_i which is not explained by the common factors and A_{ij} are the coefficients used in the combination of k factors. The general form of the evaluation of factor F_j is:

$$F_j = \sum_{i=1}^{p} W_{ji} X_i = W_{j1} X_1 + W_{j2} X_2 + \dots\dots\dots + W_{jp} X_p$$

where W_{ji} are the factor score coefficients and p is the number of variables.

The 31 items that considered affecting the perception on quality in the banking sector as described in the original BSQ model were used for measuring the perceived service quality of banks. A seven-point Likert scale, as proposed by Bahia and Nantel was used.

4 Results and Discussion

4.1 Customer Profile

The data collected were analysed according to customer use of banks in the private or the public banks. Table 1 shows the sample distribution between the two banking sectors as it relates to branch location, gender, age, and income characteristics.

It is worth noticing that mature adults above the age of 45 heavily prefer public sector banks possibly due to their past experiences and habits. The >65 years old group appears an even stronger preference for the public sector banks. Income indeed appears to affect the choice between private and public sector banks. From the sample of 519 customers that earn a monthly income between 101-300 (in 000 Drs), 201 are customers of the private sector and 318 of the public sector banks. Mainly customers of the middle and lower income classes prefer public sector banks, while on the other private banks from the middle to high-income classes.

4.2 Quality perspectives according to banking sector

Table 2 shows that, customers receive (or they perceived receiving) a different service profile. In all the 31 BSQ items, clients of private banks (Group 1) perceive receiving a higher level of quality than did those of public banks (Group 2). When applying the test on the differences between the mean scores of the two groups, it is noted that in 27 of the 31 items, the differences were statistically significant at the 1% level. Statistically the main differences between the two groups relate to the first three items of the access factor meaning, (a) "Waiting is not too long", (b) "Queues that move rapidly", and (c) "Sufficient number of open tellers", with significance $p<0.001$. Subsequently, with the use of factor analysis, we examine the main factors that formulate the quality perceptions of both private and public banks. For this purpose the factor analysis was applied as following: (a) for the total sample of 1260 private customers and (b) separately for the 525 customers of the private sector banks and (c) for the 735 customers of public sector banks.

Table 1: Characteristics of the sample (n=1260)

	Private banks	*Public banks*	*Total*
Customer use	525	735	1260
Cities	256	357	613
Region	269	378	647
Male	288	392	680
Female	237	343	580
Age			
18-24	89	137	226
25-34	137	155	292
35-44	131	151	282
45-54	97	179	276
55-65	54	64	118
> 65	17	49	66
Income (monthly in 000 Drs)			
Without income	72	113	185
< 100	16	49	65
101-300	201	318	519
301-500	128	164	292
> 500	108	91	199

Table 3 displays the results for the total sample. The reliability of the scale for each group of banks was tested using the alpha-Cronbach coefficient. As can be seen in Table 3, Principal Component Analysis and Varimax with Kaiser normalization rotation methods brought out a total of six factors, which explain 62,3% of the total variance. This percentage may not be particularly high but it is considered as quite sufficient. The evaluation of the numbers of factors was carried out using the eigenvalue criterion, which was greater, the unitary (Hair *et al.*, 1992). The analysis revealed a clear structure between factors and the loadings of the corresponding variables. The values for the variables were deemed satisfactory, particularly those of the first items of all six factors, with the higher loading price (0,863) for the "waiting is not too long". The first, most important factor explains 11,295% of the total variance, the second 11,037%, the third 10,812%, the fourth 10,417%, the fifth 9,680% and the sixth 9,050%. It should be noted that the percentage of variance of each factor could be considered as a level of importance (weight) of the relevant factor. These importance factors could have been normalised but this was judged as unnecessary since it would not differentiate the results of the survey. With regard to the reliability on the scale measurements in relation to the variables composing each factor, the alpha-Cronbach coefficients were calculated and judged satisfactory between 0,883 and 0,785.

Table 2: Mean score and *t*-tests of Private and Public Banks

Items	*Private Banks (n=525)*	*Public Banks (n=735)*	*t-test*
1. Waiting is not too long	4,72	3,65	10,591 [a]
2. Queues that move rapidly	4,87	3,89	10,123 [a]
3. Sufficient number of open tellers	4,76	3,87	8,932 [a]
4. No delays due to bureaucratic factors and procedures	4,86	4,14	7,812 [a]
5. Recognition of a regular client	5,10	4,75	4,083 [a]
6. Knowledge of the client on a personal base	4,96	4,42	5,860 [a]
7. Valorisation of the client by personnel	5,49	5,09	5,129 [a]
8. Confidentiality	5,71	5,41	4,512 [a]
9. Confidence	5,43	5,21	3,196 [a]
10. Decoration of facilities	5,61	5,04	7,329 [a]
11. Efficacious work environment	5,51	4,96	7,485 [a]
12. Cleanliness of facilities	6,13	5,76	6,195 [a]
13. Complete gamut of services	5,37	5,07	4,204 [a]
14. The range of services is consistent with the latest innovation in banking services	5,34	4,86	6,695 [a]
15. Modern equipment	5,45	5,06	5,047 [a]
16. Absence error in service delivery	5,23	4,76	6,266 [a]
17. Precision of filling systems	5,64	5,43	3,290 [a]
18. No contradiction in decisions between personnel and management	4,89	4,63	3,368 [a]
19. Delivering when promised	5,28	4,96	4,723 [a]
20. Precision of account statements	5,84	5,78	0,990
21. Well-trained personnel	5,38	4,81	7,507 [a]
22. Feeling of security	5,73	5,47	3,714 [a]
23. Good reputation	5,81	5,26	7,549 [a]
24. Sufficient number of ATMs per branch	4,88	4,63	2,722 [b]
25. Indications (communication) of quality	5,39	4,79	8,174 [a]
26. Interruption of the service	5,44	5,10	4,712 [a]
27. Reasonable fees for the administration of the accounts	4,22	4,06	2,040 [b]
28. Balance amount from which service charges begin	4,12	4,06	0,894
29. Good explanation of service fees	4,63	4,33	3,530 [a]
30. Keeping the client informed every time a better solution appears for a problem	4,70	4,33	4,240 [a]
31. The bank contacts me every time it is useful	4,87	4,48	4,339 [a]

Notes: $^{a}p<0,001$ - $^{b}p<0,005$

Table 3: Factor Analysis for all Banks (Private and Public)

Items	*Loadings*	*% of Variance*	*Grombach's a*
1. Access		11,295	0,883
1. Waiting is not too long	0,863		
2. Queues that move rapidly	0,821		
3. Sufficient number of open tellers	0,763		
4. No delays due to bureaucratic factors and procedures	0,595		
2. Effectiveness		11,037	0,832
5. Recognition of a regular client	0,802		
6. Knowledge of the client on a personal base	0,686		
7. Valorisation of the client by personnel	0,643		
8. Confidentiality	0,602		
9. Confidence	0,574		
3. Tangibles and service portfolio		10,812	0,865
10. Decoration of facilities	0,810		
11. Efficacious work environment	0,738		
12. Cleanliness of facilities	0,726		
13. Complete gamut of services	0,550		
14. The range of services is consistent with the latest innovation in banking services	0,506		
15. Modern equipment	0,494		
4. Reliability		10,417	0,792
16. Absence error in service delivery	0,693		
17. Precision of filling systems	0,679		
18. No contradiction in decisions between personnel and management	0,642		
19. Delivering when promised	0,505		
20. Precision of account statements	0,494		
21. Well-trained personnel	0,376		
5. Assurance		9,680	0,785
22. Feeling of security	0,714		
23. Good reputation	0,686		
24. Sufficient number of ATMs per branch	0,496		
25. Indications (communication) of quality	0,431		
26. Interruption of the service	0,387		
6. Price		9,050	0,805
27. Reasonable fees for the administration of the accounts	0,739		
28. Balance amount from which service charges begin	0,719		
29. Good explanation of service fees	0,592		
30. Keeping the client informed every time a better solution appears for a problem	0,563		
31. The bank contacts me every time it is useful	0,478		

The first factor refers to the ability of customers to access the bank including the relevant items. Access to the bank cashiers, appears the most important item that affects the Greek banks perceived service quality. Queuing, insufficient number of open tellers combined with delays due to beaurocratic factors and procedures, seem to constitute a big problem for the customer. According to the results, the second most important factor affecting the perceptions on service quality is the bank effectiveness. For the Greek bank customers, their recognition as regular customers meaning knowing the customer on a personal base, special treatment and valorisation by personnel is of high importance. Dealing with the customer in the manner described above, particularly if it is combined with a sense of confidentiality, is proved to be of high value too. The third factor refers to tangibles and service portfolio. Banks working environment and particularly, decoration and cleanliness of the facilities attract the customer. Certainly, a complete gamut of service is important too. Reliability, meaning absence of errors, precision in filling and generally precision in the transactions is the next most important factor. The reliability of the transactions depends mainly on the training and education of the bank's personnel. This analysis should be seen under the light of the following reality: the set of service quality dimensions has a different demanding level for customers. For this reason, in many cases a quality factor appears to have a low weight factor, and this does necessarily not mean that it is not important to customers.

Table 4: Comparison of the BSQ findings of the Canadian Bank survey with the BSQ findings of the Greek Banks survey

Rank of Factors	*BSQ*	*All Banks*	*Private Banks*	*Public Banks*
1	Effectiveness and assurance	Access	Effectiveness	Assurance
2	Access	Effectiveness	Reliability	Tangibles and service portfolio
3	Price	Tangibles and service portfolio	Access	Access
4	Tangibles	Reliability	Tangibles and service portfolio	Price
5	Service portfolio	Assurance	Assurance	Effectiveness
6	Reliability	Price	Price	Reliability

Table 4 compares the six factors of the BSQ Canadian survey with the results of the Greek survey. The comparison is possible due to the homogeneity of the sample and the use of the same instrument for conducted the research. As a limitation for the credibility of the comparison, has to be considered the fact that the sample of the Canadian survey consisted only of customers of one bank, the National Bank of Canada. In Canada, the BSQ analysis has brought up six factors that were ranked as follows: Effectiveness and assurance, Access, Price, Tangibles, Service portfolio and Reliability. The main difference between the two results is the difference in the ranking of Assurance and Price as quality factors. Canadians ranked Assurance in the first and Price in the third place respectively, while Greeks in the fifth and sixth place. Perhaps, the Greek bank customers consider their transactions more "insured" since the Greek government quarantined their deposits. The positioning of pricing as the least relevant factor to the formation of service quality perceptions, may be due to the fact that price competition is a recent development in Greek banking, that takes place at a period when interest rates and interest rate margins are low. Additionally, the differences could be due to the fact that the Canadian customer is more mature than the Greek bank customer.

The ranking also differs among the private and public sector customers. The percentage of total variance explains 62,9% for private and 61.7% for public sector banks. The ranking of the factors differs too. For the private sector banks the first factor explain the higher percentage of variation 13,1 % and 13,3%, for private and public sector bank respectively. In details, the private bank sector customers, place emphasis on effectiveness, reliability and access, while the customers of public sector bank rank the factors as following: assurance, tangibles and service portfolio and access. Price and assurance are the least important factors for both sectors respectively. The above findings combined with the customers' profile, show that the service quality perceptions differ among the two groups of customers and between the Greek and Canadian bank customers.

Comparing the BSQ findings with those of the SERVQUAL survey performed by Kagnis and Voukelatos (1997) we could argue that customers' perceptions of service quality have been differentiated. Within their survey, public banks offered convenient locations, good appearance of staff and modern banking equipment (empathy and tangibles), private banks were perceived as offering good appearance of staff, accurate statements and willingness to help customers (tangibles, reliability and assurance). Within our survey, public banks are considered providing assurance, a good environment and comparatively easy access, while private banks were perceived offering recognition and knowledge of the customer at a personal level, reliability and easy access. We can identify a "turn" on emphasis given by customers; from aesthetics to interpersonal relations. This turn may be due to the changes in the banks external environment and in particular the changing customer profile (changing consumer needs and spending patterns) that led to an emphasis on customer satisfaction through better delivery of service. Besides time and the differences in the instrument used, other important reasons for the different findings, may be the size and the location of the sample and the modernisation and upgrading of the service offered by banks that took place in the last four years.

5 Concluding Remarks

Within the context of the methodology used, the survey findings show a different profile of perceptions of service quality between the two groups of Greek Banks.

The results with respect to customer perceptions suggest that the service that is offered by banks in the private sector have a more favourable influence on actual perceptions of quality received than in the case of banks in the public sector. The evidence shows (table 2) that public sector banks do not manage the factors influencing quality as well as do the private sector banks, and hence the lower level of service quality perceived. If service quality does matter then the customers may start drifting to the private sector. The finding of this survey also suggest that banks should be looking carefully at each one of this dimensions where customers perceived receiving a different level of service than they should and address strategies to feel this gap. Access and effectiveness, the second is referring to interpersonal relations with bank's staff seem to be the most important factors that formulate perceived service quality for all banks and thus, should become the main focus of the banks marketing and human resource management strategies.

The current study appears to have some limitations. The main limitation, *as was* underlined by Bahia and Nantel, lies in the fact that the BSQ scale construction is entirely based on "expert" opinions and published literature and not on primary qualitative research with bank customers.

References

Anderson, E.W., Fornell, C. and Lehmann, D. R. (1994), "Customer satisfaction, market share, and profitability", *Journal of Marketing*, Vol. 58, July, pp. 53-66.

Arora, R. and Stoner, C. (1996), "The effect of perceived service quality and name familiarity on the service selection decision", *Journal of Services Marketing*, Vol. 10, September.

Babakus, E. and Boller, G. W. (1992), "An empirical assessment of the SERVQUAL scale", *Journal of Business Research*, Vol. 24, No. 3, pp. 253-68.

Bahia, K. and Nantel, J. (2000), "A reliable and valid measurement scale for perceived service quality of banks", *International Journal of Bank Marketing*, Vol. 18, No. 2, pp. 84-91.

Baker, J. A. and Lamp, C. W. (1993), "Measuring architectural design service quality", *Journal of Professional Services Marketing*, Vol. 10, No. 1.

Barnes, J. G., and Cumby, J. A. (1995), " The cost of service quality: extending the boundaries of accounting systems to enhance customer value", in Glyn, W. J. and Barnes, J. G. (Eds.), *Understanding Service Management*, Wiley, Chichester, pp. 178-202.

Berry, L. L., and Parasuraman, A. (1991), *Marketing Services – Competing through Quality*, Free Press, Macmillan, New York, NY.

Berry, L. L., Zeithaml, V. A. and Parasuraman, A. (1985), "Quality counts in services too", *Business Horizons*, Vol. 31, No. 5, pp. 46.

Bitner, M. J. and Hubert, A. R. (1994), "Encounter satisfaction versus overall satisfaction versus overall quality", in Rust, R. T. and Oliver, R. L. (Eds.), *Service Quality: New Directions in Theory and Practice*, Newbury Park, CA, pp. 72-74.

Bloemer, J., De Ruyter K. and Peeters, P. (1998), "Investigating drivers of bank loyalty: The complex relationship between image, service quality and satisfaction, *International Journal of Bank Marketing*, Vol. 16, No. 7, pp. 276-286.

Boom, B. H. and Britner, M. J. (1981), "Marketing Strategies and organisation structures for service firms", in Donnellly, J. and George, J. R. (Eds.), *Marketing of Services*, American Marketing Association, Chicago, IL.

Bresinger, R. P. and Lambert, D. M. (1990), "Can the SERVQUAL scale be generalized to business-to-business services?", *Knowledge Development in Marketing, AMA's Summer Educator's Conference Proceedings*, American Marketing Association, Chicago, Il, p. 289.

Capon, N., Farley, J. U., Hoening, S. (1990), "Determinants of financial performance: A meta-analysis", *Management Science*, Vol. 31, October, pp. 1143-59.

Carman, J. M (1990), "Consumer perceptions of service quality: an assessment of the SERVQUAL dimensions", *Journal of Retailing*, Vol. 66, No.1, Spring, pp. 35-55.

Caruana, A., and Pitt, L. (1997), "INTQUAL – an internal measure of service quality and the link between service quality and business performance", *European Journal of Marketing*, Vol. 31, No. 8, pp. 604-616.

Chetham J., (1994), "Issues in European Banking - Policies, strategies and practices", *Financial Times*, Financial Reports.

Cronin Jr, J. J. and Taylor, S. A. (1992), "Measuring service quality a reexamination and extension", *Journal of Marketing*, Vol. 56, July, pp. 55-68.

Curwin, J. and Slater, R. (1991), *Quantitative Methods for Business Decisions*, 3rd Ed., Chapman and Hall, London.

Davies, B., Baron, S., Gear, T. and Read, M. (1999), "Measuring and managing service quality", *Marketing Intelligence and Planning*, Vol. 17, No. 1, pp. 33-40.

De Ruyter, J. C., Bloemer, J. M. A. and Peters, P. (1997), "Merging service quality and service satisfaction: an empirical test of an integrative framework", *Journal of Economic Psychology*, Vol. 18, pp. 387-406.

Edvardsson, B. and Gustavsson, B. (1990), "Quality in services and quality in service organisations. A model for quality assessment" in Brown, S. W, Gummerson, E. Edvardsson, B. and Gustavsson, B. (Eds.), *Service Quality: Multidisciplinary and Multinational Perspective*, Lexington Books, Lexington, MA., pp. 319-339.

Edvardsson, B., Thomasson, B. and Ovretveit, J. (1994), *Quality of Service*, Barie Dale, London.

Gilmore, A and Carson, D. (1992), "Research in service quality: Have the horizons become too narrow?", *Marketing Intelligence and Planning*, Vol. 10, No. 7, pp. 5-7.

Gortsos, C. (1998), *The Greek Banking System*, Athens, Hellenic Bank Association.

Gronoos, C. (1983), *Strategic Management and Marketing in the Service Sector*, (Report, No. 83 – 104), Marketing Science Institute, Cambridge, MA.

Gronoos, C. (1990), *"Service Management and Marketing - Managing the Moments of Truth in Service Competition"*, Lexington Books, Lexington, MA.

Hair, J. F., R. E. Anderson, R. L. Tatham & W. C. Black, (1992), *"Multivariate Data Analysis, with Readings"*, Third Edition, Macmillan Publishing Company, New York.

Hallowell, R. (1996), "The relationships of customer satisfaction, customer loyalty, and profitability: An empirical study", *International Journal of Service Industry Management*, Vol. 7, No. 4, pp. 27-42.

Higgins, L. F. and Ferguson, J. M. (1991), "Practical approaches for evaluating the quality dimensions of professional accounting services", *Journal of Professional Services Marketing*, Vol. 7, No. 1, pp. 3-17.

Kagnis, P. and Voukelatos, V. (1997), "Private and public banks: a comparison of customer expectations and perceptions", *International Journal of Bank Marketing*, Vol. 17, No. 5, pp. 279-287.

Kano, N., N. Seraku, F. Takahashi and S. Tsuji (1996*)*, Attractive quality and must be quality, in: J. D. Hromi (Ed.), *The best on quality*, International Academy for Quality, The Quality Press, Milwaukee, WI, 7.

Kloutsiniotis, A.G. (1995), " Strategic choices in the development of Greek banks", *Economic Review of the Commercial Bank of Greece*, Vol. 1, pp. 4-22.

Kosmidou, K., and C. Spathis (2000), "Euro and Profitability of Greek Banks", *European Research Studies Journal*, (in press).

Lassar, M. W., Manolis, C. and Winsor R. (2000), "Service quality and satisfaction in private banking", *The International Journal of Bank Marketing*, Vol. 18, No. 4, pp. 181-99.

Mihelis, G., E. Grigoroudis, Y. Siskos, Y. Politis and Y. Malandrakis (2001), "Customer satisfaction measurement in the private bank sector", *European Journal of Operational Research*, Vol. 130, 2, pp. 347-360.

Moschos, D. and Fragetis, D. (1997), The present and future of Greek banks, *Current issues*, IOBE, Athens.

Noulas, A. (1999), "Profitability and Efficiency of the Greek Banks (1993-1998)", *Journal of the Banking Association of Greece*, 19/20, 53-60.

Palmer A. and Cole, C. (1995), *Services Marketing: Principles and Practice*, Prentice-Hall, Englewood, Cliffs, NJ.

Parasuraman, A., Zeithaml, V. A. and Berry, L. L. (1985), "A conceptual model of service quality and its implications for future research", *Journal of Marketing*, Vol. 49, April, pp. 41-50.

Parasuraman, A., Zeithaml, V. A. and Berry, L. L. (1988), "SERVQUAL: a multiple-item scale for measuring consumer perceptions of service quality", *Journal of Retailing,* Vol. 64, Spring, pp. 12-40.

Petridou, E., and Glaveli, N. (2000), "The changing banking environment and developments in banks training effort: The case of Greece", *presented in AT.IN.E.R.'s 2nd International Conference – Business Economics, Marketing, Management: Past, Present and Future,* Athens, June.

Pruden, D. R., R. Sankar and T. G. Vavra (1996), "Customer loyalty: the competitive edge beyond satisfaction", *Quirk's Marketing Research Review,* Vol. 24, 49-53.

Reichheld, F. and Sasser, W. E. (1990), "Zero defections" *Harvard Business Review,* pp. 105-11.

Richard, M. D. and Allaway, A. W. (1993), "Service quality attributes and choice behavior", *Journal of Services Marketing,* Vol. 7, No.1, pp. 59-68.

Rust, R. T., Zahoric, A. J. and Keiningham, T. L. (1994), "Return on Quality (ROQ), making service quality financially accountable", *Journal of Marketing,* Vol. 59, pp. 58-70.

Spathis, C., Kosmidou, K. and Doumpos, M. (2001), "Assessing profitability factors in the Greek banking system: A multictriteria methodology", *presented in 10th IFORS Special Conference - New Trends in Banking Management,* Athens, April.

Spreng, R. A. and Mckoy, R. D. (1996), "An empirical examination of a model of perceived service quality and satisfaction", *Journal of Retailing,* Vol. 72, No. 2, pp. 201-14.

Taylor, S. and Baker, T. (1994), "An assessment of the relationship between service quality and customer satisfaction in the formation of customers' purchase intentions", *Journal of Retailing,* Vol. 70, No. 2, pp. 163-78.

Vavra T. G. (1995), *Aftermarketing: How to keep customers for life through relationship marketing* , Irwin Professional Publishing, Burr Ridge, IL.

Vavra, T. G. (1997), *Improving the measurement of customer satisfaction,* ASQC Quality Press, Milwaukee.

Zahoric, A. J. and Rust, R. T. (1992), "Modeling the impact of service quality on profitability: a review" in Swartz,T. A., Bowen D. E. and Brown, L. L. (Eds.), *Advances in Service Marketing and Management,* Vol. 1, JAI Press, Greenwich, CT, pp. 247-76.

Zeithaml, V. A. and Britner, J. M. (1996), *Services Marketing,* McGraw-Hill, New York, NY.

Zeithaml, V. A., Parasuraman, A. and Berry, L. L. (1990), *Delivering Quality Service – Balancing Customer Perceptions and Expectations,* The Free Press, New York, N.Y.

Zikmund, G. W., (1994), *"Business Research Methods",* Fourth Edition, The Dryden Press, N.Y.

Zopounidis, C., D. K. Despotis and E. Stavropoulou (1995), "Multiattribute Evaluation of Greek Banking Performance", *Applied Stochastic Models and Analysis*, 11, 1, 97-107.

Zopounidis, C., M. Doumpos and H. Georgiou (1997), "Application of a Multicreteria Methodology in the Assessment of Banking Branches Efficiency", *Economic Review of the Commercial Bank of Greece,* 10, pp. 4-9.

Assessing Efficiency of Bank Branches: An Application of Data Envelopment Analysis (DEA)

A. Anastasiou, J. Mitropoulos, A. Sissouras

University of Patras
Department of Management and Operational Research
Rio 26500 Patras, Greece

Summary:
In recent years, considerable emphasis has been given to the determination of the efficiency of public and private organisations and service units using the Data Envelopment Analysis Method (DEA). The method evaluates the relative efficiency of the units under scrutiny. It compares branches objectively to identify the best-practice branches, the less productive branches, and the changes the less productive branches need to make in order to reach the best operating level and to improve their profitability. Bank management finds the DEA results provide meaningful insights not available from other techniques that focus on ways to improve productivity whilist the ability to model both quantitative and qualitative factors in its structure extends the usefulness of the method. This study is concerned with the assessment of the efficiency from a sample of thirty-nine bank branches of the "Commercial Bank" of Greece. The results have evidenced the disparities in efficiency which exist between the branches, whilst specific efficiency targets are determined for the inefficient branches. The results suggest that DEA is a beneficial management tool in the field of rational decision making and strategic planning.

Keywords: Efficiency, Data Envelopment Analysis, Commercial Bank of Greece, Greece

1 Introduction

Commercial banking is based on the operation of networks of bank branches which act as the key contact point between the customers and the central bank. Bank branches facilitate the banking production process via the attraction of various types of deposits that generate sufficient assets to be invested by the bank. From the marketing point of view bank branches are service providers that reflect the commitment of the corporate bank to understand customers' needs and provide products and services of high quality. Banks operate under increasing pressure to improve the operating efficiency of their branches. The operating efficiency isn't constrained within a typical 'cost minimisation' framework but it is given a broader dimension which covers issues related to the attraction of customer, cost management, cash intermediation and finally quality of services. These four aspects constitute the performance portfolio of bank branches which contribute to the overall performance of the corporate bank (Athanassopoulos, 1997).

The primary objective of measuring and evaluating the operating efficiency of bank branches in a competitive environment is not only to position the branches with respect to each other in terms of their efficiencies, but also to gain insight into the nature of operations so that managerial measures can be taken to improve their performance. It is necessary for management to identify and develop means of improving branch performance. For this purpose, other bank management tools that compensate for the weaknesses in accounting and financial ratios are needed. It seems that Data Envelopment Analysis (DEA) is such an approach.

The purpose of this paper is to describe an application of DEA to performance assessment of 39 commercial branches of a large banking network in Greece, the "Commercial Bank of Greece". The related data come from the Commercial Bank statement for the year 1999 and cover the regions of Peloponissos and Western Greece. The experience gained during this empirical study indicates that DEA can be considered as an alternative bank management tool to traditional accounting and financial ratios since it offers means of more comprehensively assessing the operating efficiency of bank branches.

In that stage it would be useful a brief account to be done on the developments of the basic financial magnitudes of the Commercial Bank within the four years period of 1997 to 1999. During that period, the profitability was increased gradually, but at a quick pace, the capital position was tripled, the main credit magnitudes of the Bank were extended significantly and in general terms its shares were hold to the basic magnitudes of the banking market within a period of intense rearrangements. The competitiveness index was improved substantially, the number of personnel was confined, and the share price of the Bank, which is the mirror of the above developments, went up satisfactorily and reached higher levels than the increase levels in the index prices of the banking shares.

Also, at the end of 1999 the network of the Bank's branches included 380 branches, covering geographically the whole country. The branches are characterized by their size, which is typically indicated by the number of employees. Branches with fewer than five employees are considered small, those with six to 19 employees, medium, and those with 20 to 40 are considered large to very large. Regional offices have up to 155 employees. The Commercial Bank and its subsidiary companies are activated in nine foreign countries. The increase rate of the number of branches, ATMs and personnel are depicted in Figure 1.

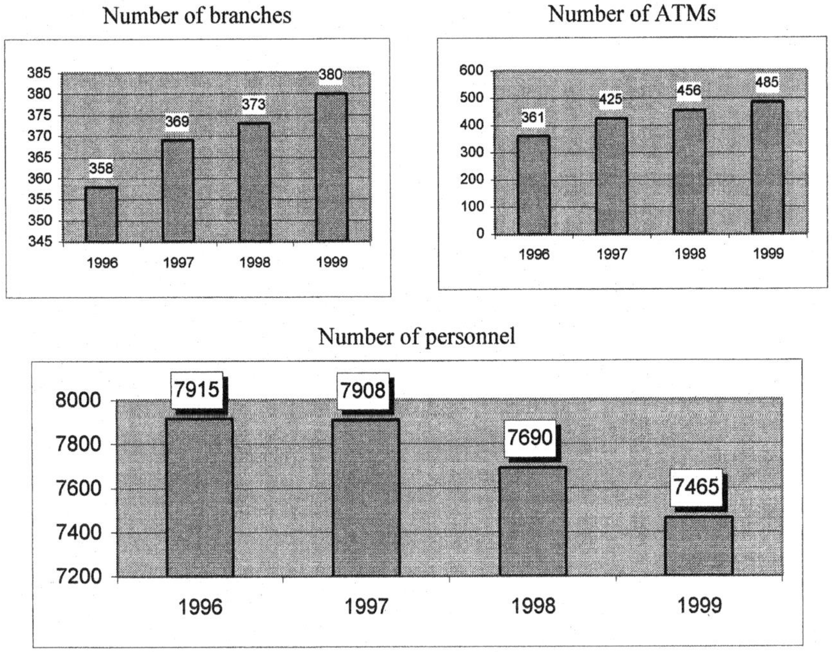

Figure 1: Number of branches, ATMs and personnel for the period of1996 – 1999

Afterwards, in 1999 the Commercial Bank promoted successfully banking products and services in the area of middle businesses and private individuals. In the market of businesses the Bank corresponded successfully to its 50000 customers' needs. According to the Bank's system of market management it expanded 900 new co-operations by a number of 75 billion drachmas drachmas (1GBP≈550 DRACHMAS)in financings, changing, as a whole, its grantings to middle businesses at the amount of 700 billion drachmas (increase percentage 19%).

2 Methodology Framework

The basic prerequisite of all the methods of productivity measurement is the systemic approach of the productive process, which is obtained by an organization through an multiple input-output system. The term " output " represents the products or services produced by the units.The term " input ", for the main part, represents the resources used (human and material) in producing the output and, secondarily, as the environmental factors which influence the formation of the resources (e.g. state of organization, coverage, location etc.).

The method centers on the determination of the relatively efficient units, i.e., those units which demonstrate the most successful input-output transformation (successful is defined as that transformation which, while maintaining minimum input, obtains maximum output or while maintaining maximum output, obtains minimum input). These units constitute the referrals "standards" and "envelop" the other units, thus producing the 'efficient frontier' as is shown in Figure 2.

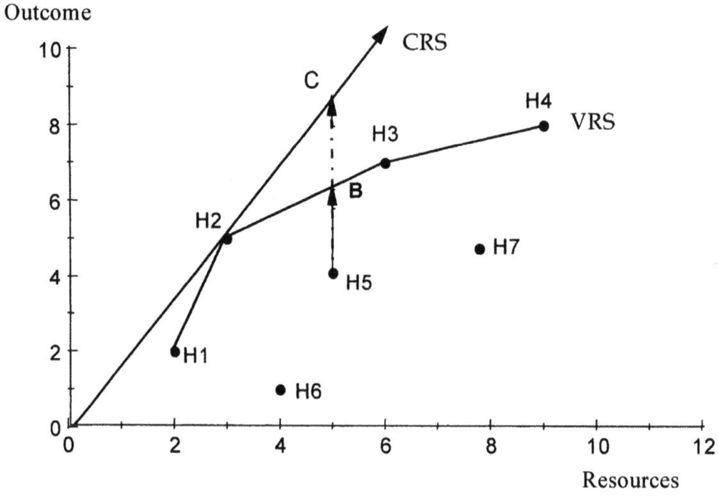

Figure 2: Branches Profiling Against Frontier

Figure 2 exhibits the input (resources) and output (outcome) levels of a number of operating branches H1 to H7.The attempt to assess the performance of individual branch has comparative nature and thus one has to define a standard upon which this assessment will be performed. Branch H2 has the highest level of outcome per unit of resources and thus defines an efficient frontier under constant returns to scale (CRS) that passes through the origin.Relaxing the earlier assumption one can also define an efficient frontier under variable returns to scale (VRS) that is made by branches H1, H2, H3 and H4.The sales efficiency of individual branch is made by projecting them on the efficient frontier (e.g. H5B under VRS and H5C under CRS for branch H5). The relative efficiency of each branch can be obtained by solving the optimisation models in [i] or [ii] under constant and variable returns to scale respectively.An instance of these models for branch k is given in [i] and [ii] below.

[i] CRS – Model	[ii]VRS - Model
$\min_{\theta,\lambda} \theta$,	$\min_{\theta,\lambda} \theta$,
St. $-y_i + Y\lambda \geq 0$,	St $-y_i + Y\lambda \geq 0$,
$\theta x_i - X\lambda \geq 0$	$\theta x_i - X\lambda \geq 0$,
$\lambda \geq 0$,	$N1'\lambda = 1$
	$\lambda \geq 0$,

From the above equations it is assumed that there are data on K inputs and M outputs for each of N units. For the i-th unit these are represented by the column vectors x_i and y_i, respectively. The K×N input matrix, X, and the M×N output matrix, Y, represent the data for all N units. In (i), θ is a scalar and λ is a N×1 vector of constants. This envelopment form involves fewer constraints than the multiplier form (Coelli et al., 1999), hence it is generally the preferred form to solve. The value of θ obtained will be the efficiency score for the i-th unit. It will satisfy θ≤1, with a value of 1 indicating a point on the frontier hence a technically efficient unit. Note that the linear programming problem must be solved N times, once for each unitin the sample. A value of θ is then obtained for each unit. In (ii) N1 is an N×1 vector of ones. This approach forms a convex hull of intersecting planes which envelope the data points more tightly than the CRS conical hull and thus provides technical efficiency scores which are greater than or equal to those obtained using the CRS model. The VRS specification has been the most commonly used specification in the 1990s.

The efficiency of branch n is expressed in percentage terms as E_n^{CRS} and E_n^{VRS} respectively.The mechanism for assessing the efficiency of each branch can be summarised as follows:

- One separate linear programming problem is solved for each branch within the sample .
- The empirical frontier is defined using non-dominated branches i.e. no other branch or combination of branches can be found that with the same or less resources can generate more output,
- Inefficient branches are projected on the efficient frontier adopting orientations such as output maximisation (for given resources maximise outputs), input minimisation (for given outputs minimise resource use) or mixed strategies.

The frontier based assessment of branches performance has profound implications that go beyond the strict measurement of relative efficiency (Athanassopoulos, 1997).These can be summarised as follows:

- **Search for the unique**. The assessment of the performance is undertaken seeking to identify the best match between the operating characteristics of the assessed branches, say n, and combination of efficient branches that will be used as its benchmarks.The criterion of Pareto-dominance is applied to identify those branches that qualify as comparators (peers) of inefficient branches.

- **Efficiency profiling.**For inefficient branches the assessment yields a summary measure of their distance from their position on the efficiency frontier.A league table is provided, therefore, regarding the performance of the branches within the industry concerned.The efficiency rating is obtained from the objective function of [i] and [ii].
- **Target projections**. The frontier profiling mechanism apart from the relative efficiency values yields also target positions for inefficient branches according to a projection formulae which shows the extent to which inputs / outputs need improving in comparison to their benchmarks.
- **Economic assumptions.**Frontier analysis models are flexible in encapsulating alternative returns to scale assumptions between inputs and outputs.That is, the technical efficiency of individual branches can be assessed under the assumptions of constant, non-increasing, non-decreasing and variable returns to scale.The formulation listed in [i] and [ii] contains the two most widely used models that are associated with the assumptions of constant and variable returns to scale.

3 Literature Review

Historically the D.E.A. method started with the use of benchmarking and target setting as proposed by Farrell in 1957, but was developed more systematically after 1978 by Charnes, Cooper and Rhodes (1978). Since then, rapid progress has followed the initial model, not only with respect to the theory but also to its applications. More specifically in banking sector during the late 1980s and particularly in the 1990s, the DEA method has been used extensively to evaluate banking institutions. However, only a few studies focused on the analysis of bank branches within the same financial institution. Published applications of DEA to bank branches are summarised in Table 1. This table contains information about the authors, the year of publication, the type of inputs and outputs, the country and sample size.

The relation between efficiency and profits was first analyzed by Oral and Yohalan (1990). They defined two DEA models for analyzing both efficiency and profitability. The profitability model consisted of a desegregation of expenses and income, which were considered as inputs and outputs, respectively. However, the implications of using this model instead of a usual profitability measure, such as profitability ratio, were not discussed.

Table 1: Published applications of DEA to bank branches

Author and year	Inputs	Outputs	Sample & country
Zenios et al. (1999)	Managerial personnel	No. of current personal accounts	144 bank branches
	Clerical personnel	No. of savings accounts	from
	Computer terminals	No. of foreign currency and commercial accounts	Bank of Cyprus
	Work space	No. of credit applications	
		Ouput time	
AthanassopoulosA	No. of employees	No of deposit accounts	68 bank
(1997)	No. ATMs and teller machines	No. of credit transaction	branches
	No. of computers terminals	No. of debit transactions	of Greek
		No. of loan applications evaluated	Comercial Bank
		No. transactions involving commissions	
Drake L,	No. of interview rooms	Till transactions	190 bank
Howcroft B	No. of ATMs	Lending products	branches
(1994)	Square meters of branch space	Deposit products	from a
	Management grades	Automated transfers	UK bank
	Clerical grades	Clearing items	
	Stationary costs	Ancillary business	
		Insurance business	
Oral M,	No. of personnel	Time spent on general	20 bank
Yohalan R	No. of on-line terminals	service transactions	branches
(1990)	No. of commercial accounts	Time spent on credit transactions	from a Turkish
	No. of saving accounts	Time spent on deposit transactions	bank
		Time spent on foreign exchange transactions	
Sherman HD, Gold F	No.full-time equivalent employees	No. more resource consuming transactions	14 bank branches
(1985)	Rent paid	No.medium/high resource consuming trans.	from a US Savings
	Total cost of supplies	No. medium/low resource consuming trans.	Bank

4 Input - Output Selection

The correct definition of the inputs and outputs for bank branches is not straightforward and controversy remains in the literature, giving rise to alternative approaches. The use of three resources (inputs) and four services (outputs) in this study was based on management's assessment of key branch activities as denoted in previous section (Literature Review). The aim of this efficiency assessment was the minimisation of disposable resources in order to produce a steady size level of inputs. This means that the DEA model will examine the branch cost efficiency and endeavour to minimise the resources (Table 2).

Table 2: Input-Output specification for branch efficiency

Inputs	Outputs[1]
I1. Number of Managerial Personnel	O1. Total Commissions (in thousand drachmas)
I2. Number of Clerical Personnel	O2. Total Grantings (in thousand drachmas)
I3. Total Branch Cost (in thousand drachmas)	O3. Value of Current Accounts (in thousand drachmas)
	O4. Value of Savings Accounts (in thousand drachmas)

On the hand, the inputs used in the DEA model are: The number of managerial personnel, the number of clerical personnel and the total branch cost. The choice of the first two inputs represents the resource commitments of individual branches from the personnel point of view (Athanassopoulos, 1997; Camanho and Dyson, 1999; Oral and Yolalan, 1990; Soteriou and Zenios, 1999a; Zenios et al., 1999) and the third input, total branch cost, includes the rent, the operational cost and the labour cost (Drake and Howcroft, 1994; Kantor and Maital, 1999; Sherman and Gold, 1985; Soteriou and Zenios, 1999b). Moreover, the number of ATMs and the floor space of the branch were not included in the analysis due to data unavailability. Note that the technology within the branch hasn't been considered in the model, as the branch equipment is fairly homogeneous in this network. Hence, variations in efficiency are not attributed to technological differences, but rather on cost disparities concerning resources (human & material).

On the other hand, as observed from the relevant bibliography (Athanassopoulos, 1997; Camanho and Dyson, 1999; Kantor and Maital, 1999; Soteriou and Zenios, 1999a; Thanassoulis, 1999) four output dimensions are used in the assessment of production efficiency: total commissions, total grantings, value of current accounts and value of savings account. The first output, consists

[1] 1GBP≈550 DRACHMAS

the combined private banking commissions, import-export commissions and credit card commissions. If assumed that these transactions are similar enough in terms of resources used on average to be treated as one group of transactions. For the same reason, total grantings consist the number of real estate loans, the number of commercial loans and the number of personal loans. The four outputs span all the activities covered by individual branches avoiding aggregation of heterogeneous types of transactions. An assumption that all four outputs incur the same degree of complexity to bank employees is implicitly adopted if the assessment of branch efficiency were based on the raw number of transactions for different activities. This assumption is relaxed in this paper incorporating value judgments that reflect the differential degree of complexity between bank branch operations. Other service transactions excluded by the model cause of insignificant volume (such as purchase and sale of foreign currency) or of bank didn't have adequate records to include the service (such as withdrawals and traveller checks).

5 Empirical Results

According with relevant bibliography (Athanassopoulos, 1997; Drake and Howcroft, 1994; Sherman and Gold, 1985), the use of the dual form of the CRS DEA model (I) considered the most appropriate. The results of the DEA analysis are depicted in Table 3 and indicate that the 33% of the entire sample has efficient branches and the 67% has inefficient branches. That is there is scope for efficiency improvements in 26 branches.

As reported in Table 3, based on the data provided on the outputs and inputs, DEA identified 26 of the 39 branches to the relatively inefficient. Each branch with an efficiency rating of less than 1.0 is relatively inefficient. The efficiency rating does not rank order the branches, but rather suggests the degree of inefficiency, of a branch compared with its efficiency reference set. For example, Branch 31 is about 75% efficient compared to BR1, BR21, BR22, BR28 and BR36. The Branch 39 isabout 77% efficient, but compared with BR10, BR21, BR22, BR25, BR28 and BR36. Generally, this means that both Branch 31 and Branch 39 should be able to reduce each of their input levels by about 25% and 23% respectively without reducing their output levels to be as efficient as their respective efficiency reference sets. Similarly, Branch 2 has the lowest efficiency rating — 33% — which means it would have to reduce its inputs by 67% to become as efficient as its reference set ofBR21, BR22, BR25, BR28 and BR36.

Table 3: The DEA results

Branch code	DEA efficiency rating	Efficiency reference set (peers)
BR1	1.0	
BR2	0.328	BR21, BR22, BR25, BR28, BR36
BR3	0.586	BR10, BR 36
BR4	0.649	BR1,BR10,BR21,BR22,BR28,BR36
BR5	1.0	
BR6	0.908	BR1, BR21, BR35
BR7	1.0	
BR8	0.740	BR21, BR22, BR28, BR36
BR9	1.0	
BR10	1.0	
BR11	0.595	BR21, BR28, BR36
BR12	0.579	BR10, BR25, BR28, BR30
BR13	0.723	BR22, BR28
BR14	0.583	BR1, BR7, BR22, BR25, BR28
BR15	0.838	BR1,BR10,BR21,BR22,BR28,BR36
BR16	0.948	BR21, BR22, BR25, BR28
BR17	0.474	BR1,BR10,BR21,BR22,BR28,BR36
BR18	0.949	BR21, BR25, BR28, BR30
BR19	0.539	BR1, BR21, BR22, BR28, BR36
BR20	0.762	BR21, BR22, BR25, BR28
BR21	1.0	
BR22	1.0	
BR23	0.787	BR10, BR21, BR28, BR36
BR24	0.750	BR21, BR28
BR25	1.0	
BR26	0.663	BR1, BR25, BR36
BR27	0.519	BR1, BR21, BR22, BR35
BR28	1.0	
BR29	0.776	BR21, BR22, BR25, BR28
BR30	1.0	
BR31	0.750	BR1, BR21, BR22, BR28, BR36
BR32	0.427	BR1, BR21, BR22, BR28, BR36
BR33	0.912	BR1, BR21, BR35
BR34	0.546	BR1, BR21, BR22, BR28
BR35	1.0	
BR36	1.0	
BR37	1.0	
BR38	0.903	BR21, BR22, BR35
BR39	0.773	BR10,BR21,BR22,BR25,BR28,BR36

Moreover, Branch 3 is less efficient with a DEA efficiency rating of 58.6%, suggesting that it could provide its current mix and volume of services with only about 58.6% of the resources it actually consumed. The Branch 12 has a DEA efficiency rating of 57.9% indicating that it is using about 42.1% excess resources. The analysis shows that branches such as Branch 1 and Branch 7 are efficient branches, indicated by the DEA efficiency rating of 100%. In fact, 26 of the 39 branches are using excess resources and 20 branches are using 20% or more excess resources.

These findings indicate that the Commercial Bank can make substantial efficiency improvements. The results also indicate the inefficient branches and their efficiency reference set (Table 3). For example, the Branch 15 is inefficient specifically in comparison to BR1, BR10, BR21, BR22, BR28 and BR36. These reference sets (peers) indicate the branches that are almost similar to the relatively inefficient branches in their mix of services and resources, and they are used to identify the specific operating characteristics that make the relatively inefficient branch more costly to operate.

Furthermore a detailed branch analysis can be performed. For example, Bank Branch 24 which is identified as inefficient, with efficiency rating, θ=0.75. As can be seen from Table 3, the efficiency reference set of Branch 24 consists of Bank Branch 21 and Bank Branch 28. The 'Composite Branch', which is the hypothetical efficient branch that Bank Branch 24 would like to become, can be defined in terms of these two efficient branches. Recalling that the Composite Branch is a linear combination, in which the coefficients are the dual variables λ_{Bj}, of the members of the efficiency reference set, it can be written

$x_{iC} = \lambda_{B21}\, x_{i21} + \lambda_{B28}\, x_{i28}$ [iii]

for input i, where x_{iC} is the amount of input i needed at the Composite Branch C. Put differently, x_{iC}'s are the amounts of inputs that Bank Branch 24 should have been using to produce the same amounts of outputs currently being produced in order to be considered as efficient. More specifically, the amounts of input Bank Branch 24 should have been using are

$$x_{1C} = (0.563)\,(2) + (0.469)\,(4) = 3.002,$$
$$x_{2C} = (0.563)\,(6) + (0.469)\,(4) = 5.254,$$
$$x_{3C} = (0.563)\,(70137) + (0.469)\,(124583) = 97916.558.$$

To clearly see the sources of the inefficiency of Bank Branch 24, Table 4 is prepared by using the formulas in [iii] with λ_{B21} =0.563 and λ_{B28} =0.469.

Table 4: Excess use of inputs by Bank Branch 24

Inputs	Composite branch (1)	Branch 24 actual (2)	Excess use of inputs (2) – (1)
x_1=managerial pers.	3.002	4	0.998
x_2=clerical pers.	5.254	7	1.746
x_3=total branch cost	97916.558	143440	45523.442

It is clear from the third column of Table 4 that Bank Branch 24 is currently using more of every kind of input compared to the Composite Branch. Take for instance, the total branch cost actually used and the total branch cost suggested by the DEA model. The difference of 45523 thousand drachmas about being identified as excess total branch cost in a branch of 143440 thousand drachmas is rather significant by any standards.

A similar argument is used in determining the number of managerial and clerical personnel to be employed in each branch. Also, a similar argument, like the detailed analysis of Bank Branch 24, is used for each branch in order the excess use of inputs to be determined. Thus, it was done that analysis for all the branches and the results are depicted in Table 5.

Table 5: Excess inputs based on DEA

Inefficient	DEA	I_1	%Decrease	I2	%Decrease	I3	%Decrease
BR1	1	2	0	5	0	68151	0
BR2	0.328	2	73.1	3	67.2	46275	67.2
BR3	0.586	2	41.3	3	41.3	53976	46.3
BR4	0.649	2	35	3	35	54462	35
BR5	1	2	0	5	0	68622	0
BR6	0.908	7	9	14	9.4	228464	16.3
BR7	1	9	0	7	0	180849	0
BR8	0.740	15	26.5	17	26	351904	26
BR9	1	5	0	10	0	175331	0
BR10	1	2	0	3	0	64739	0
BR11	0.595	2	56.1	2	40.5	46543	40.5
BR12	0.579	2	50.4	2	42.1	53634	42.1
BR13	0.723	2	58.5	1	27.6	29748	27.6
BR14	0.583	10	41.7	10	41.6	273416	45.3
BR15	0.838	12	16.21	1	16.2	270632	16.2
BR16	0.948	7	5.2	7	5.2	174400	5.2
BR17	0.474	17	52.3	19	52.3	386213	52.3
BR18	0.949	2	53.5	1	5.8	51298	44
BR19	0.539	4	46	9	46	129569	46
BR20	0.762	12	23.8	13	23.8	319770	23.8
BR21	1	2	0	6	0	70137	0
BR22	1	9	0	6	0	158311	0
BR23	0.787	3	21.3	5	21.3	106337	30.7
BR24	0.750	4	24.9	7	24.9	143440	31.7
BR25	1	4	0	5	0	124336	0
BR26	0.663	2	33.6	3	34.7	51669	33.7
BR27	0.519	6	48	10	48	185363	55
BR28	1	4	0	4	0	124583	0

Table 5: Excess inputs based on DEA (*continued*)

Inefficient	DEA	I_1	%Decrease	I2	%Decrease	I3	%Decrease
BR29	0.776	11	23.7	12	22.3	241927	22.3
BR30	1	3	0	3	0	90705	0
BR31	0.750	5	24.9	6	24.9	115966	24.9
BR32	0.427	4	57.3	5	57.3	90444	57.3
BR33	0.912	16	8.8	33	8.8	587822	25.7
BR34	0.546	4	45.5	6	45.5	107089	49.2
BR35	1	9	0	11	0	219954	0
BR36	1	2	0	3	0	46333	0
BR37	1	2	0	5	0	73157	0
BR38	0.903	5	9.7	9	9.7	157896	22.5
BR39	0.773	10	22.7	13	22.7	258286	22.7

Analysing the above results it can be stated that the mean overall efficiency of the branches is 0,795%, pointing that serious change by organisation's recourse management can be established in the field of strategic programming and planning. More specifically 25% of branches seems to posses low efficiency profile (<0,595) while a rest 25% of breaches posses medium efficiency rate (0.595 – 0,787). The analytic statistical results are depicted in the following table.

Table 6: Analysis of efficiency rating (E^{CRS})

ECRS

N	**Valid**	***39***
	Missing	***0***
Mean		***,795051***
Std. Error of Mean		***3,20E-02***
Mode		***1,0000***
Std. Deviation		***,200129***
Minimum		***,3280***
Percentiles	**25**	***,595000***
	50	***,787000***
	75	***1,000000***

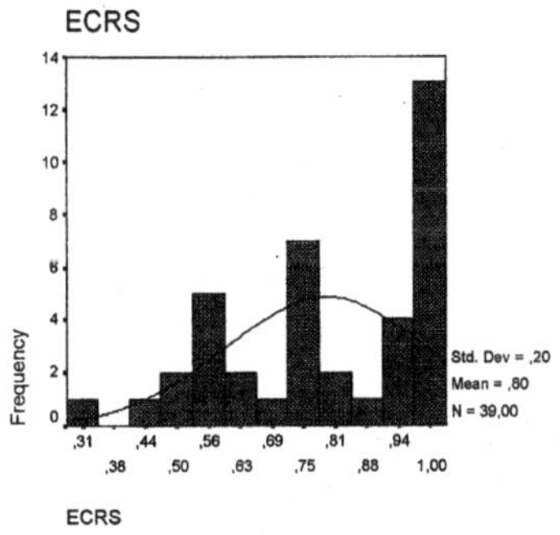

From table 3 it can be seen also that the Branch 21 appears more frequently (21) than any other branch and the Branch 7 appears only once. However, the fact that the Branch 21 appears 21 times shows that this is the ideal branch. Thus, these 21 inefficient branches, which are compared with Branch 21, should have almost the same magnitudes with that branch. The magnitudes of the Branch 21 are: the number of managerial personnel is two, the number of clerical personnel is six and the total branch cost is 70.137 thousand drachmas. Therefore from a policy point of view the external and internal environment of branch 21 must further be examined in order to established general targets of bank brunch productivity.

To sum up, the Bank uses the model in many capacities. It establishies policy guidelines based on the model's recommendations. Policy guidelines take two forms. If management perceives through the analysis of the model that a branch has an "ideal" mix of accounts but is inefficient, it recommends reducing resources (mainly personnel). No layoffs are involved: The bank usually transfers personnel to other branches that would benefit from inputs. Branches that are perceived to be inefficient because of an inappropriate mix of products receive the model results along with guidelines on how to achieve specific targets, including the specific structure of the accounts. The Bank's marketing department uses this information to better focus on the kinds of accounts that would improve the mix (Zenios et al., 1999).

6 Summary - Discussion

Traditionally, banks used to measure the efficiency of their branches separately at the strategic and tactical levels. In strategic settings, they defined branch efficiency by simple operational ratios (for example, transactions per teller) or by financial ratios (for example, deposits to loans or return on assets). At the tactical level, they used industrial- engineering methods to measure time-and-motion efficiency (that is, micromanaging teller operations and optimising work procedures or queuing systems). As a common practice they measured each of these aspects separately. Today, on the other hand, banks generally realize that simultaneous assessment, incorporating multiple quantitative and qualitative data sources, is necessary to evaluate branch operations (Golany and Storbeck, 1999).

Consistent with the above logic the present study focuses on the efficiency of branch cost and production in comparison with the movement of branch capital. Thus, the Data Envelopment Analysis model was used to identify cost and production efficient branches and estimates allocative cost-minimising efficiency. The application of this method to 39bank branchesaimed:

A. To identify the overall efficiency profile in branch level and at the same time, ascertain those branches with low or medium productivity.
B. To point out to the non-efficient branches their reference model(s) (peers) and establish the particular intervention targets with regards to the functioning and programming of their services.
C. To assist the organization's recourse management in an attempt to achieve economies in scale with optimization in the use of resources.

Summarising, it can be stated (Zenios et al., 1999) that continued improvements in a Bank's operations are essential in a competitive environment. Such improvements are, however, increasingly difficult to bring about, and they require novel and sophisticated modeling techniques. Data envelopment analysis is a powerful tool for analysing the efficiency of bank branches; it can consider the external environment that is outside managerial control, the microenvironment that the Bank has created through its long-term efforts and that it can alter in the long run as well, and the internal environment that managers can alter directly and immediately. Not only does the model identify sources of ineffieciency, but it also provides constructive recommendations for improvement. Furtheremore, the model can benchmark the effects of external factors, thus providing useful information for strategic planning.

References

Athanassopoulos, A.D. (1997), "Service quality and operating efficiency synergies for management control in the provision of financial services: Evidence from Greek bank branches", *European Journal of Operational Research* **98**, 300-313.

Camanho, A.S. and Dyson, R.G. (1999), "Efficiency, size, benchmarks and targets for bank branches: an application of data envelopment analysis", *Journal of the Operational Research Society* **50**, 903-915.

Charnes, A., Cooper, W.W. and Rhodes, E. (1978), "Measuring the efficiency of decision making units", *European Journal of Operational Research* **6**(2), 429-444

Coelli, T., Raw, P. and Pattese, G. (1999), *An introduction to efficiency and productivity analysis*, Kluwer Academic Publishers, London.

Drake L. and Howcroft B. (1994), "Relative efficiency in the branch network of UK Bank: an empirical study", *Omega* **19**:83-90.

Golany, B. and Storbeck, J.E. (1999), "A Data Envelopment Analysis of theOperational Efficiency of Bank Branches", *Interfaces* **29**, 14-26.

Kantor, J. and Maital, S. (1999), "Measuring Efficiency by Product Group: Integrating DEA with Activity-Based Accounting in a Large Mideast Bank", *Interfaces* **29**, 27-36.

Oral, M. and Yolalan, R. (1990), "An empirical study on measuring operating efficiency and profitability of bank branches", *European Journal of Operational research* **46**, 282-294.

Sherman, H.D. and Gold, F. (1985), "Bank Branch Operating Efficiency", *Journal of Banking and Finance* **9**, 297-315.

Soteriou, A. and Zenios, S.A. (1999), "Operations, Quality, and Profitability in the Provision of Banking Services", *Management Science* **45**(9), 1221-1238.

Soteriou, A. and Zenios, S.A. (1999), "Using data envelopment analysis for costing bank products", *European Journal of Operational Research* **114**(2), 234-248.

Thanassoulis, E. (1999), "Data Envelopment Analysis and Its Use in Banking", *Interfaces* **29**, 1-13.

Zenios, C.V., Zenios, S.A., Agathocleous, K. and Soteriou, A.C. (1999), "Benchmarks of the Efficiency of Bank Branches", *Interfaces* **29**, 37-51.

Comparative Satisfaction Measurement of ATMs vs Tellers

A. Bouranta[1], A. Kouremenos[1], Y. Siskos[2]

[1] University of Piraeus
80, Karaoli & Dimitriou
18534 Piraeus, GREECE

[2] Technical University of Crete
University Campus
73100 Chania, GREECE

Summary:
Measuring customer satisfaction is the most reliable quality index and it should help firms to continuously improve their provided services or products. The intense competition that characterizes the banking system (amalgamations, redemptions, creation new banking institutions) comes to emphasize the importance of customer satisfaction measurement. This paper presents an original pilot bank customer satisfaction survey dealing with the evaluation of quality of services provided by ATMs and tellers in Greece.

Keywords: Automated Teller Machine, Teller, Multicriteria Analysis, Customer Satisfaction Analysis

1 Introduction

Nowadays, it has become clear to firms that it costs less to keep their customers than to increase them. Relative surveys have indicated that it costs about five times more to gain a new customer than to keep an existing one (Parasuraman et al., 1985). Measuring customer satisfaction is a necessary tool for banking marketing (Gerson, 1993, Hill, 1996). It is a common knowledge that customer satisfaction mainly depends on the process of service delivery and less on the outcome of it. The quality of provided services is a customer judgement and results from the discrepancy between customers' expectations and perceptions (Stevens, 1995, Peters, 1991, Crosby 1979, Garvin, 1983, Lovelock, 1983).

The objective of this paper is the comparison of service quality bank customers received by ATMs and tellers by evaluating the corresponding customer satisfaction (Anderson et al., 1986, Bahia and Nantel, 2000). It is an original survey in the Greek bank sector in which took part bank customer over 18 years old. The sample was quite small and was selected by the statistical method of snowball.

The paper is organized into five sections. Section 2 emphasizes briefly the definition of the two concepts: ATM and teller. Section 3 presents the basic principles of the system MUSA as well as the implemented methodological frame. The main results of the survey are given in section 4 while summarizes some concluding remarks and proposals. Finally, a brief presentation of the MUSA method is provided in appendix.

2 Definition of ATM nd Teller

Automated Teller Machines or ATMs are a part of the new banking technology (Electronic Banking). They are capable of doing a variety of transactions without the support of human cashiers (Direct Banking). These machines operate with a magnetic stripe on the back of plastic cards and with the personal identification number typed in by the customer. They can offer a 24 hour-service, every day of the year. Barclays Bank in the UK claims to have installed the first Cash Dispenser in the world at Enfield in June 1967. National Bank of Greece installed the first ATM in Greece approximately in 1992. At present, there are about 2.800 ATMs in Greece in operation (Goode and Moutinho, 1995, Moutinho and Smith, 2000, European Central Bank, 1999).

On the other hand, tellers interact with the majority of customers (80%) and generally handle a wide range of banking transactions. The variety of activities performed by tellers started changing since the beginning of the ‘80s. That has occurred because a lot of people don’t perceive any more the branches like the point of their banking activities. Electronic banking gives them the opportunity to do easy transactions away from the teller windows. So, tellers are free from performing simple, standardized financial services and enrich their roles with more complex functions. In the following years, the percentage of specialized employees, is expected to rise whereas the search for employees with more increased skills and quality characteristics (e.i selling and communication skills, responsibility, initiative) will be more intense. In a few words, there is going to be a shift of employees from the productive line towards sales and customer support (Lovelock, 1996).

3 Methodological Frame

The MUSA (Multicriteria Satisfaction Analysis) system of Grigoroudis and Siskos has been used in order to measure customer satisfaction (Appendix). The basic principle of MUSA is the aggregation of individual judgements into a collective value function, assuming that customer 's global satisfaction depends on a set of criteria representing service characteristic dimensions (Roy, 1985, Siskos et al., 1997, 1998, Mihelis et al., 2001) (figure 1). According to the survey, each customer is asked to express his/her own judgements, namely his/her global satisfaction and his/her satisfaction with regard to a set of discrete criteria. Based on these assumptions, the problem is approached as a problem of qualitative regression and solved via special linear programming formulations where the sum of deviations between the customer 's global satisfaction evaluation and its multicriteria satisfaction evaluations is minimized (Siskos and Yannakopoulos, 1985, Ankiran, 1994, Cronin and Taylor, 1992, Gnanadesikan, 1977). With the use of MUSA system one can estimate the customer global satisfaction by ATMs and tellers, their satisfaction according to the criteria and subcriteria as well as the weights of these criteria. Combining weights and satisfaction indices, the improvement and customer loyalty diagrams can be developed.

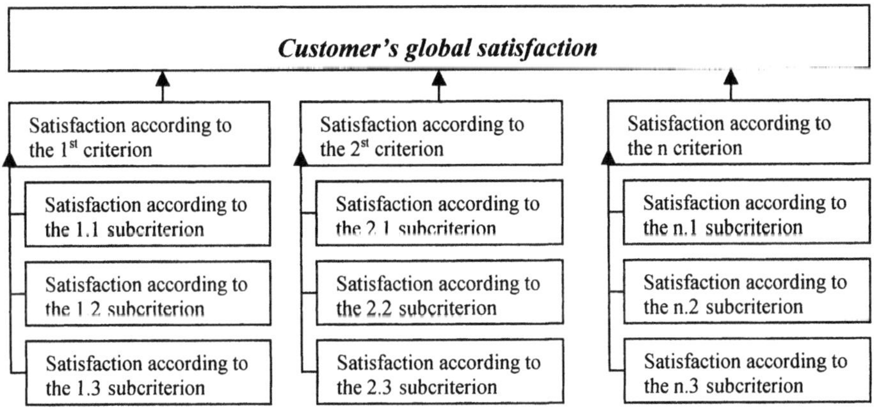

Figure 1: Aggregation of customer's preferences

Approximately, 100 clients of retail Greek banks participated in this pilot survey. The sample was selected according to the snowball (Churchill, 1999, Goodman, 1961). In many particular applications, nonprobabilistic methods like snowball are used for selecting samples, primarily having as a first target convenience. Otherwise, the derivation of unbiased point estimates and confidence intervals with specified probability content could not be achieved with strict statistical validity. Input data were collected using a very simple and short questionnaire of 14 questions. Some of the data are quantitative and some of them are qualitative. Five-point Likert scales (highly dissatisfied to highly satisfied) were used to measure expectations and perceptions.

4 Basic Results

4.1 Preliminary Analysis

According to the surveys of Parasuraman every customer identifies with five general dimensions which represent the evaluation criteria he/she uses to assess the quality of provided services (Parasuraman et al., 1990, Gronroos, 1982,). The main satisfaction criteria consist of:

Reliability: Ability to perform the promised service dependably and accurately.

Responsiveness: Willingness to help customers and to provide prompt service.

Empathy: The ability to provide caring, individualized attention to its customers.

Tangibles: Appearance of physical facilities, equipment, personal and communication materials.

Assurance: Courtesy of employees and their ability to convey trust and confidence.

The hierarchical structures of the set of criteria and subcriteria about ATMs and tellers are presented in figures 2 and 3.

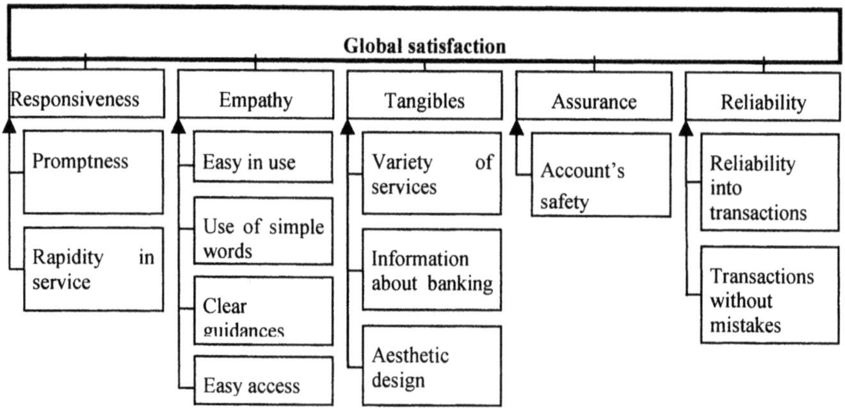

Figure 2: Hierarchical structure of customer 's satisfaction dimensions of ATMs

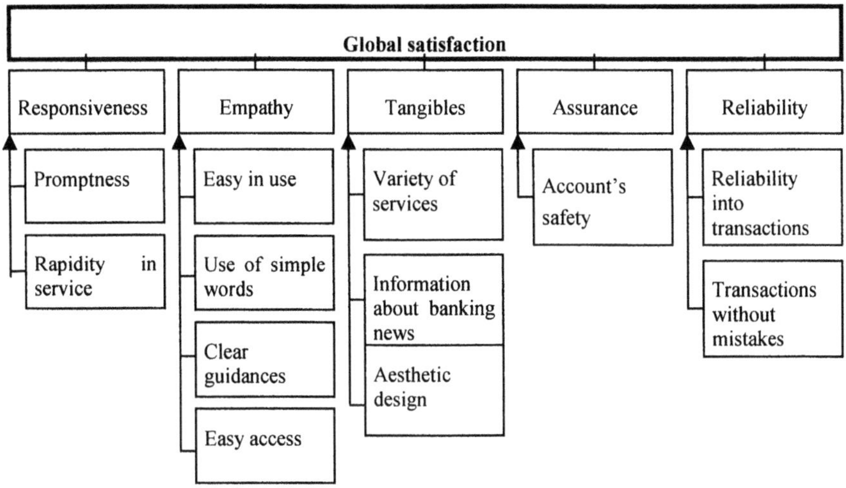

Figure 3: Hierarchical structure of customer 's satisfaction dimensions of tellers

4.2 Satisfaction Analysis

The customer satisfaction analysis referring to ATMs and tellers is presented in figures 4 and 5. It is found that a very high percentage of customers (69,1%) seems to be satisfied of ATMs and only 30,9% are disappointed with them. So, 7 out of 10 Greek ATM users state that they are satisfied, whereas the rest seem to have experienced or to have been informed of an event, which was the main excuse for forming a bad opinion. On the other hand, the majority of bank customers are dissatisfied (59%) from their cooperation with tellers and only 41% are satisfied. As a result the banks have to focus on improving the teller 's service quality since an important space for improvement is being noticed in that area.

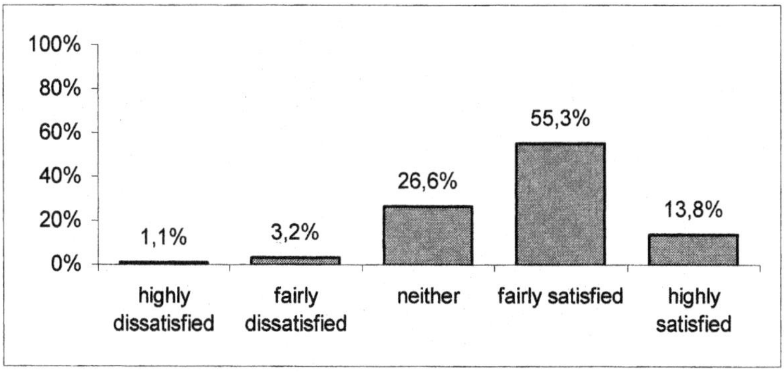

Figure 4: Global satisfaction concerning ATMs

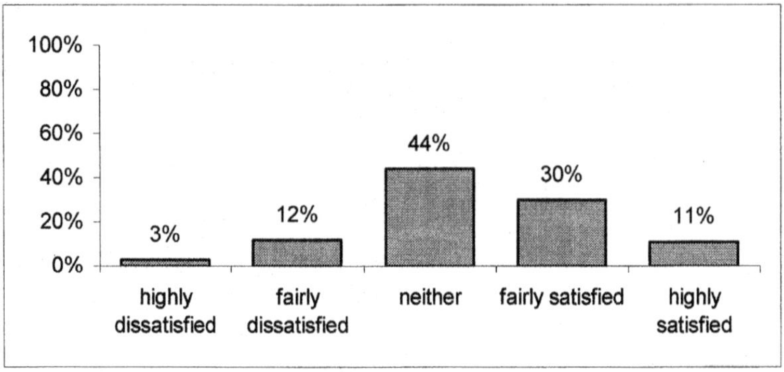

Figure 5: Global satisfaction concerning tellers

It should be noted that the average global satisfaction index from ATMs is 13% higher than the corresponding index from tellers. More specifically, this index referring to ATMs has a very high value (78,9%) in contradiction to the same index concerning tellers (65,9%). The 13% difference gives a clear message that customers receive higher service quality by ATMs than by tellers' windows. Moreover, criteria satisfaction analysis concerning ATMs, shows that customers are quite satisfied as far as reliability, responsiveness and empathy are concerned, while lower satisfaction index appears concerning tangibles (60,3%) (figure 6). The most important criteria seem to be empathy and reliability (figure 7). This can justify the high value of the global satisfaction index.

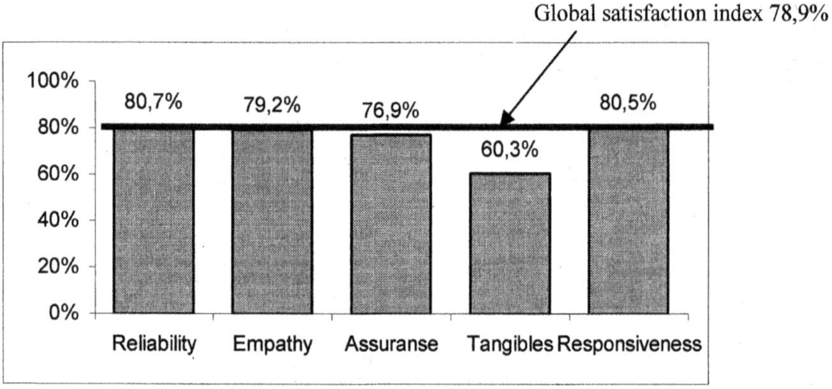

Figure 6: Average satisfaction indices concerning ATMs

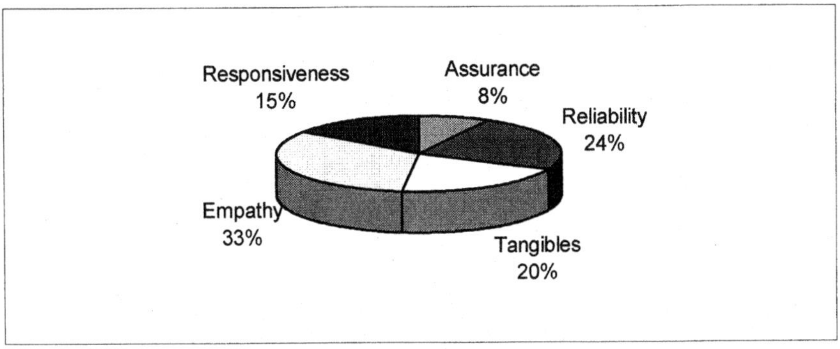

Figure 7: Criteria weights concerning ATMs

Customers seem to be less satisfied with the service provided by tellers, while the total set of criteria moves around low indices of satisfaction (68%-40,9%). The criterion tangibles has an extremely low satisfaction index of 40,9% (figure 8). The most important criterion seems to be responsiveness with a significant weight of 27% and the criterion empathy seems to play a less important role on customer's preferences (figure 9).

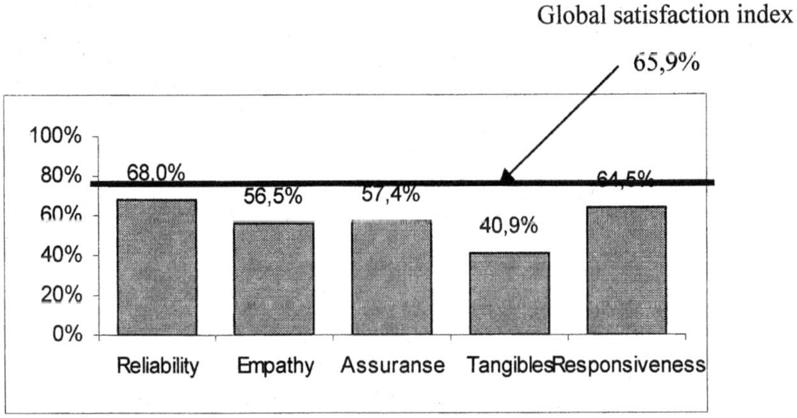

Figure 8: Average satisfaction indices concerning tellers

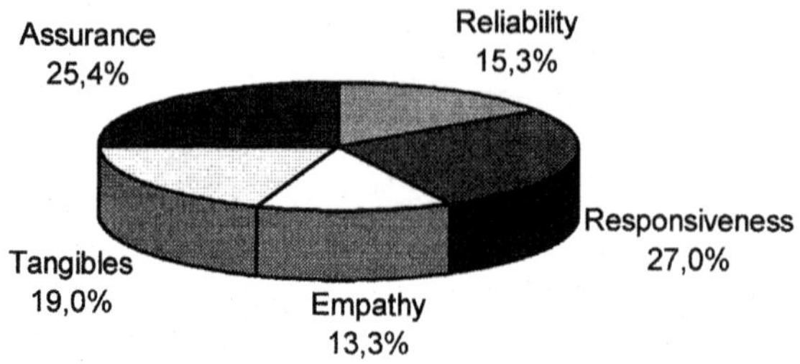

Figure 9: Criteria weights concerning tellers

It is important to notice that all quality criteria according to ATMs have higher value than the same criteria about tellers. Nevertheless, it is obvious that customers tend to give importance to different criteria concerning ATMs and tellers.

At the level of subcriteria, the main results of MUSA method are presented in Tables 1 and 2:

Table 1: Subcriteria weights and satisfaction indices for ATMs

Subcriterion	Weight (%)	Satisfaction index (%)
Reliability into transactions	6,6	73,6
Transactions without mistakes	16,9	87,9
Easy in use	8,3	80,3
Use of simple words	8,3	78,2
Clear guidances	8,3	78
Easy access	8,3	80,3
Account's safety	8,3	76,9
Variety of services	7,0	65
Information about banking news	6,1	50,8
Aesthetic	6,6	65,2

Table 2: Subcriteria weight and satisfaction indices for tellers

Subcriterion	Weight (%)	Satisfaction index (%)
Reliability	7,1	66,8
Responsibility	8,1	69,2
Comminication in easy terms	6,9	64,8
Solve individualized problems	6,5	48,1
Good personal relationship	6,4	58,5
Secrecy	6,3	59,7
Courtesy	6,5	61,1
Patience	6,2	50,4
Knowledge of banking services	6,2	62,0
General	6,5	58,3

design		
Rapidity in service	8,3	86,9
Promptness	7,0	74,2

financial knowledge		
Personal appearance	6,5	64,4
Resolving the problem quicly	6,1	57,6
Rapidity in service	6,5	57,3
Service promptness	14,2	78,6

Based to the previous results, the following observations are being made (Table 1 and 2):

- The majority of customers haven't experienced any negative events as far as transactions on ATMs are concerned, such as withholding cards and wrong money measurement. As a result, the subcriterion transactions without mistakes has the highest satisfaction index (16,9%) and at the same time is the most important service dimension (87,9%).
- ATM manufactures should try to improve the aesthetic design of ATM (i.e offer a color ATM) and probably have to continue their efforts because of the low value (65,2%) of the subcriterion aesthetic design. On the other hand, they shouldnot pay great attention to making ATMs friendlier to the users because the subcriteria easy in use (80,3%), use of simple words (78,2%) and clear guidances (78%) have a high satisfaction index.
- Knowledge of banking services or products (62%) and general financial knowledge (57,3%) have a quite low satisfaction index. Tellers should participate in sessions designed to improve their knowledge of both new and existing retail products and services as well as their financial education.
- Customers seem to be more satisfied with ATMs (86,9%) according to service dimension rapidity in service than with the same service by tellers (57,3%). Maybe, this difference is one of the main reasons for making transactions on ATMs.

Segmentation analysis has also been performed in order to identify the distinctive critical groups of customers and analyze their opinions:

- Customers who express their opinions without having ever used ATMs say that they are not satisfied with them (48%). On the contrary, frequent users (more than twice a week) are satisfied (82%) whereas they are not so mush satisfied with tellers (66%).
- Young customers (age 18-28) are more satisfied (79,8%) with ATMs than the older ones (age 51+) (69,9%). Young people accept easier the new technology than the older ones who are at the beginning more reserve. The customers who are 51 and over years old are not very satisfied with tellers (66,7%) so it is a matter of time to accept and use ATMs more frequently.
- Customers who do complex transactions on ATMs pay more attention to reliability (17,6%) and assurance (11,3%).

- Customers who visit the banks more than twice a week in order to do their transactions are more satisfied with ATMs than with tellers. Nevertheless, this specific group of customers seems to pay attention to the subcriterion politeness with which customers are quite satisfied (78,7%).

Performance/ Importance diagrams, produced by the MUSA software, indicate the critical satisfaction dimensions which should be improved and should be the priority of improvement efforts (Customers Satisfaction Council, 1995). There are no criteria located in the critical quadrant; consequently empathy, reliability as well as assurance should be the first priorities for improvement because there is room for improvement (figure 10).

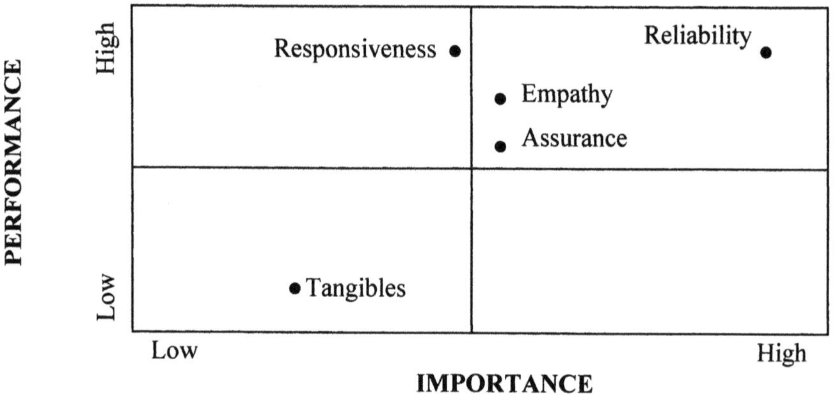

Figure 10: Performance/ Importance diagram for ATMs

Figure 11 presents the Performance/Imprortance diagram for tellers. Empathy, assurance and tangibles, are not terrible pertinent at the time of this analysis, they may be more importantin the future, and bank's performance is relatively good.

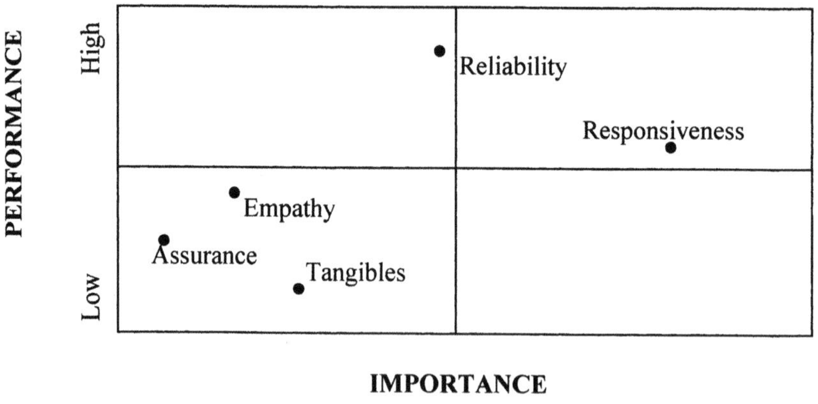

Figure 11: Performance/Importance diagram for tellers

5 Concluding Remarks

The most important results of this survey could be summarized on the following points:

- The average global satisfaction index with ATMs is 13% higher than the corresponding index with tellers. More specifically, this index concerning ATMs has a very high value (78,9%) in contradition to the same index concerning tellers (65,9%).
- 70% of the customers seem to be quite satisfied with ATMs, whereas only 41% are satisfied with tellers.
- Transactions without mistakes is the most important service dimension (16,9%) with the highest satisfaction index (87,9%).
- The service promptness arise as a competitive advantage of tellers (weight:78,6% and satisfaction index.
- Young customers (age 18-28) are more satisfied with ATMs than the older ones (age 51+).
- The frequent users (more than twice a week) of ATMS state their satisfaction in using them, as opposed to those who express their opinion without having ever used them.

The results of this satisfaction research should be used by marketing managers of banks in order to improve the quality of given services or products by ATMs and tellers. More specifically, the following actions are being proposed:

- Banks have to induce their customers to use ATMs in order to do simple transactions and to go to the tellers windows for more complex activities.
- 17% of bank customers has never used ATMs. On the other hand, the users of ATMs don't know and have never used less commont services of ATMs as stock exchange transaction, personalise and dispense cheques. So, banks should communicate (advertising spots, promotion and brochures) the services of ATMs, the manner of its operations as well as the advantages of its use.
- They must improve the service quality provided by tellers. Particularly, they must improve their service according to the subcriteria rapidity in service (57,3%), patience (50,4%), willingness to solve individualized problems (48,1%), subcriteria that move around low indices of satisfaction.

It should be noted that customer satisfaction is a dynamic parameter because of the new evolution of technology, the creation and input of new products or services and changes in to competitive field and in social facts. As a result the satisfaction barometer is considered necessary.

Moreover, it is worth noting that this exploratory pilot survey was based on a comparatively small and perhaps unrepresentative sample. While the results are intuitively appealing, further work will be necessary to replicate these findings with a large sample of bank customers.

The philosophy and the methodological frame which is used at this research should be used at any case of measuring customer satisfactions because it is simple, cheap and gives reliable results.

References

Anderson T., Cox P. and Fulcher G. (1986), "*Bank selection decisions and market segmentation*", Journal of Marketing, Vol.40, pp.40-45.

Ankiran N. K. (1994), "*Developing an istrument to measure customer service quality in branch banking*" Vol.12, pp.10-18.

Bahia K. and Nantel J. (2000), "*A reliable and valid measurement scale for the perceived service of banks*", The International Journal of Bank Marketing" Vol.2, pp.84-91.

Churchill G. (1999), "*Marketing Research: Methodological Foundations*", Dryden Press, U.S.A.

Cronin J. and Taylor A. (1992), "*Measurement service quality: a re-examination and extension*", Journal of Marketing, Vol.7, pp.55-68.

Crosby P. (1979), "*Quality is free: The art of Making Quality Certain*", New American Library, New York.

Customers Satisfaction Council (1995), "*Customer Satisfaction Assessment Guide*", Motorola, University Press.

European Central Bank (1999) "*The effects of technology on the EU banking systems*", July 1999.

Garvin D. (1983), "*Quality on the Line*", Harvard Business Review, September-October 1983, pp.65-73.

Gerson F. (1993), "*Measuring customer satisfaction : a guide to managing quality service*" , Crisp Publications, Menlo Prak.

Gnanadesikan R. (1977), "*Methods for statistical data of multivariate observations*", Wiley, New York.

Goode M. and Moutinho (1995), "*The effects of free banking on overall satisfaction: the use of automated teller machines*", The International Journal of Bank Marketing, Vol.13, pp.33-44.

Goodman L. (1961) "Snowball Sampling", Annuals of Mathematical Statistics 32, pp.168-170.

Gronroos C. (1982), "*Strategic Management and Marketing in the Service Sector*", Swedish School of Economics and Business Administration, Helsingfors.

Hill N. (1996), "*Handbook of customer satisfaction measurement*", Gower Publishing, Hampshire.

Lovelock C. (1983), "*Classifying Services to Gain Strategic Marketing Insights*", Journal of Marketing, Summer 1983, pp.9-20.

Lovelock C. (1996), "*Services Marketing*", Present Hall International, USA.

Mihelis G., E. Grigoroudis, Y. Siskos, Y. Politis and Y. Malandrakis (2001) "*Customer satisfaction measurement in the private bank sector*", European Journal of Operational Research, 130, (2), pp.347-360.

Moutinho L. and Smith A. (2000), "*Modeling bank customer satisfaction through mediation of attitudes towards human and automated banking*", The International Journal of Bank Marketing, January 2000, pp.124-132.

Parasuraman A., Berry L. and Zeithaml V. (1985), "*A conceptual model of service quality and its implications for future research*", Journal of Marketing, Vol 49, pp.41-50.

Parasuraman A., L. Berry and A. Zeithaml (1990), "*Delivery quality service*", The Free Press.

Peters T. (1991), "*Total quality leadership: Let's get it right*", Journal for quality and participation, Vol.3, pp.10-15.

Roy B. (1985), "*Méthodologie multicritere d' aide à la Décision*", Economica, Paris.

Siskos Y., E. Grigoroudis, C. Zopounidis and O. Saurais (1998), "*Measuring customer satisfaction using a collective preference disaggregation*" Journal of Global Optimization Vol.12, pp.175-195.

Siskos Y. and D. Yannacopoulos (1985), "*UTASTAR : An ordinal regression method for building additive value functions*", Investigaçao Operacional, Vol.5, pp.39-53.

Siskos Y., E. Grigoroudis, C. Zopounidis and O. Saurais (1997), "*A Multicriteria methodology to assess customer satisfaction*", Management, Vol.1, pp.39-43.

Stevens P. (1995), "*Dineserv: a tool for measuring service quality in restaurants*", The Cornell, H.R.A Quartely, Vol. 36, pp.56-60.

Appendix: A Brief Presentation of the Musa Method

The preference disaggregation model MUSA (MUlticriteria Satisfaction Analysis) assesses global and partial satisfaction functions Y^* and X_i^* respectively, given customers' judgements Y and X_i. Following the principles of ordinal regression analysis under constraints using linear programming techniques, the ordinal regression analysis equation has as follows:

$$\begin{cases} \widetilde{Y}^* = \sum_{i=1}^{n} b_i X_i^* - \sigma^+ + \sigma^- \\ \sum_{i=1}^{n} b_i = 1 \end{cases} \tag{1}$$

where $\widetilde{Y}^*$ is the estimation of the global satisfaction function Y^*, X_i^* are the partial satisfaction functions, σ^+ and σ^- are the overestimation and the underestimation error, respectively, and b_i is the weight of the i-th criterion.

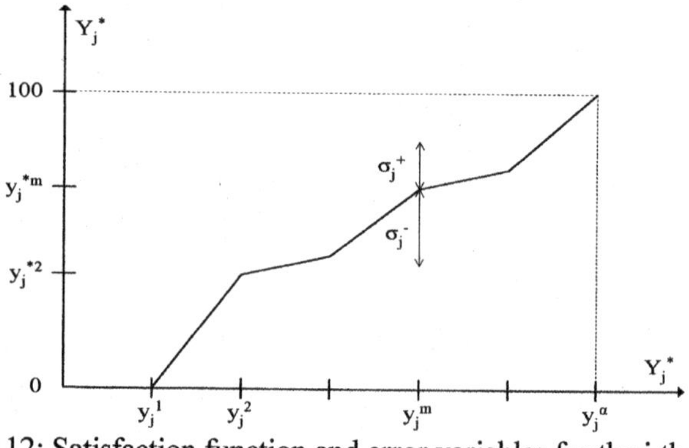

Figure 12: Satisfaction function and error variables for the j-th customer

It should be noted that Y^* and X_i^* are monotonic functions normalised between 0 and 100 (see Figure 12). Also, in order to reduce the number of the mathematical constraints the following transformation equations are used:

$$\begin{cases} z_m = y^{m+1} - y^m & \text{for m} = 1, 2, \dots, \alpha - 1 \\ w_{ik} = b_i x_i^{*k+1} - b_i x_i^{*k} & \text{for k} = 1, 2, \dots, \alpha_i - 1 \text{ and i} = 1, 2, \dots, \text{n} \end{cases} \tag{2}$$

According to the aforementioned definitions and the assumptions, the basic estimation model can be written in a linear program formulation, as it follows:

$$\begin{cases} [\min]F = \sum_{j=1}^{M} \sigma_j^+ + \sigma_j^- \\ \text{under the constraints} \\ \sum_{i=1}^{n} \sum_{k=1}^{x_i^j - 1} w_{ik} - \sum_{m=1}^{y^j - 1} z_m - \sigma_j^+ + \sigma_j^- = 0 \quad \text{for j} = 1, 2, \dots, \text{M} \\ \sum_{m=1}^{\alpha - 1} z_m = 100 \\ \sum_{i=1}^{n} \sum_{k=1}^{\alpha_i - 1} w_{ik} = 100 \\ z_m \geq 0, w_{ik} \geq 0 \quad \forall \text{ m, i and k} \\ \sigma_j^+ \geq 0, \sigma_j^- \geq 0 \quad \text{for j} = 1, 2, \dots, \text{M} \end{cases} \tag{3}$$

where M is the number of customers, n is the number of criteria, and x_i^{*j}, y^{*j} are the j-th level on which variables X_i and Y are estimated.

The preference disaggregation methodology consists also of a post optimality analysis stage in order to face the problem of model stability. The final solution is obtained by exploring the polyhedron of near optimal solutions, which is generated by the constraints of the above linear program. This solution is calculated by n linear programs (equal to the number of criteria) of the following form:

$$\begin{cases} [\max]F' = \sum_{k=1}^{\alpha_i - 1} w_{ik} \quad \text{for i} = 1, 2, \dots, \text{n} \\ \text{under the constraints} \\ \text{F} \leq \text{F}^* + \varepsilon \\ \text{all the constraints of LP (3)} \end{cases} \tag{4}$$

where ε is a small percentage of F^*. The average of the solutions given by the n LPs (4) may be taken as the final solution. In case of non-stability this average solution is less representative.

The assessment of a performance norm may be very useful in customer satisfaction analysis. The average global and partial satisfaction indices are used for this purpose and can be assessed according to the following equations:

$$S = \sum_{m=1}^{\alpha} p^m y^{*m} \text{ and } s_i = \sum_{k=1}^{\alpha_i} p_i^k x_i^{*k} \tag{5}$$

where S and s_i are the average global and partial satisfaction indices, and p^m and p_i^k are the frequencies of customers belonging to the y^m and x_i^k satisfaction level, respectively.

A Multicriteria Approach to Assess Banking Performance: The Case of Greece

O. Stathas[1], K. Kosmidou[2], M. Doumpos[2], C. Zopounidis[2]

[1] Economic and Business Management Department
Mediterranean Agronomic Institute of Chania
73100 Chania, Greece

[2] Technical University of Crete
Department of Production Engineering and Management
Financial Engineering Laboratory
University Campus, 73100 Chania, Greece

Summary:
The Greek banking system in the last years has become a system of thoroughly changes. In this new environment the Greek commercial banks should be adapted and develop new strategy and inclinations, in order to satisfy depositor demands and increase their security and profitability. A multicriteria approach is proposed for evaluating banking performance over multiple criteria. Firstly, a multivariate statistical analysis is applied to select the most important financial ratios, which derive from the study of financial statements of a sample of Greek commercial banks. Then, the multicriteria method PROMETHEE is applied to rank the banks according to their financial performance during the period 1995-1999.

Keywords: Banking performance, financial ratio analysis, multivariate statistical analysis, multicriteria decision aid.

1 Introduction

During the last years the banking sector has been characterized by continuous organizational, institutional and technological changes. The major changes in the Greek banking system, realized after 1992, include the interest rate liberalization, the annulment of various credit rules, the deregulation of capital movements and the free entrance of banking institutions into the European Union (Noulas, 1999). Nowadays, the Greek banking system is characterized by a high degree of concentration and competitiveness. Moreover, the changeover towards the monetary union and the new technological innovations herald major changes in the banking environment and challenge all banks to make timely preparations in order to enter into the new competitive monetary and financial environment.

The complexity of the banking institutional framework and the expected changes, due to worldwide vulnerable economic conditions and new tactics and tendencies, have compelled experts and researchers to continuously investigate into further specialized methods for the evaluation of the diversification of competitive banking institutions.

Therefore, it is interesting to investigate the performance of Greek banks in the new competitive and financial environment that is modulated in the view of the European Monetary Union and the adoption of the common EURO currency.

The purpose of this study is to present some financial aspects of the Greek banking system and apply a multicriteria decision aid methodology in order to evaluate banking performance. The analysis is conducted on 20 Greek banks operating in the Greek market during the period 1995-1999. The evaluation of their performance is based on their financial characteristics. For these purpose five groups of financial ratios are selected including liquidity, investment activity, capital structure, profitability and development ratios. These financial ratios represent the major areas of interest in bank analysis.

The paper is organized as follows: Section 2 presents a literature review of the Greek banking system and the methodologies that have been applied on the evaluation of the banking performance. Section 3 describes the methodology employed in this study, while Section 4 presents the data used, the application and the obtained results. Finally, in section 5 the concluding remarks are discussed.

2 Literature Review

2.1 The Greek Banking System

The number of banks in Greece increased substantially during the second half of the nineteenth century (Gortsos, 1998). After a considerable consolidation of the Greek banking sector in the late 1890s, several new banks were established during the first decade of the 20th century. The 1920s was a period of major changes in the structure of the Greek banking system. Although conditions did not favor the establishment of new banks, since the 1960s the internationalization of banking activities led to an increase in the number of foreign banks opening branch offices in Greece. This trend was reinforced by the association of Greece with the European Community in the 1970s and then the membership of Greece in the European Community in 1981.

In 1982, and especially after 1987, the Greek financial system underwent a series of intense changes. Allocation inefficiencies, initiation of the process for financial integration within the European Union, and international trends towards globalization and deregulation, all contributed to the rise of a program modernizing the Greek financial system and adapting it to internationally acceptable standards.

Nowadays, the Greek banking system in its effort to prepare itself for the changeover to the EURO, faces some additional challenges. Banks must follow a strategic plan and prepare themselves to enter the new competitive monetary and financial environment (Hardouvelis, 1997). This will have an impact on business, systems and delivery. A radical change on the whole banking system is necessary in order to adopt the new currency and cooperate with other European banks (Vasiliou, 1993; Kloutsiniotis, 1996; Garganas, 1998; Karamouzis, 1998). It is obvious that during the last decade, there is a restructurization of bank operations within all segments of the banking business. These major changes increased the existing competitiveness, and denoted the need for reforms through mergers and privatization (Tolios, 1998). To examine the costs that arise from this changeover and the benefits that are produced, Kosmidou and Spathis (2000) estimated the change in the bank profits. This analysis showed that during the period 2002-2007, banks will face a decrease in their profits. Further analysis indicates that profits will rapidly increase in the long-term period. In view of the EURO, banks will try to recover from the loss of revenues that will be caused following the introduction of the EURO by looking for new income resources. This loss and the eminent competition urge banks towards the purchase of small or medium businesses aiming at a creative relationship with the customer - entrepreneur. Apart from loans and advisory consulting, this purchase includes investment banking, access to international financial markets, as well as alternative ways of capital movements through leasing, factoring and venture capital. Zopounidis *et al.* (1995) dealt with the illustration of an ordinal utility model upon a sample of Greek commercial banks for the period 1989-1992, in order to evaluate their banking performance over multiple criteria.

The liberalization and the profitability of the Greek commercial banks during the years 1989-1991 were examined by Alexakis *et al.* (1995). The results suggest that the determinants of profitability of Greek commercial banks are highly different from those depicted in other countries during the periods of intense regulation in Greece. The cost structure and the scale economies in the Greek banking system during the years 1980-89 were examined by Karafolas and Mantakas (1996). Variables such as size of assets, capital, labor and technological progress were used. By exploiting the properties of the model, they were able to show whether technological progress was the factor that affected banking costs. Papaioannou and Ganzonas (1997) examined the development perspectives of non financial products for the Greek Banking system. Zopounidis *et al.* (1997) evaluated the efficiency of the banking branches of the Bank of Greece using a multicriteria decision aid classification method. Three classification models were developed classifying the branches into three efficiency groups with classification accuracy of more than 85%.

Hardy and Simigiannis (1998) examined the competitiveness and the effectiveness of the Greek banking system. They realized that during the decade of 1990s only few banks succeeded to maintain stable high levels of profitability.

This was done especially by medium sized banks that were not state-controlled and could keep a certain rate of return.

Spathis and Kosmidou (1999) examined the competitiveness measured by Tobin's Q ratio between Greek small and large banks during the period 1990-1998. The evidence indicated that the competitiveness of Greek banks depends on their efficiency in asset/liability management, the current assets to loan ratio and the size as measured by the total assets.

As far as the profitability and the efficiency of Greek banks are concerned, Noulas (1999) examined the ROE and ROA ratios as well as ratios of leverage and operating efficiency for 19 Greek banks for the period 1993-1998. According to the results there are no significant differences in the return on equity and asset diachronically. Bank profitability was not better in 1998 than in 1993 or in 1994. The profitability of banks during the last two years (1997-1998) seems to increase, though, when compared to that of 1996. This year is representative, as few banks and especially the state-controlled ones, in their effort to improve their portfolios and to show reduced accounting profitability, have kept large amounts in the provisions account.

Staikouras and Steliaros (1999) examined the profitability factors of 17 commercial Greek banks for the years 1991-1998. They used the ROE and ROA ratios in relation to endogenous and exogenous variables. According to the results, the profitability of Greek banks is affected by the inflation rate, the proprietary regime, the ratio of reserve funds for borrowings to the total of granted loans and the ratio of debts to the total assets. Hondrogiannis *et al.* (1999) examined the competitive conditions of the Greek banking system for the period 1993-1995. The results indicate that bank revenues were earned under the conditions of monopolistic competition. This gradual elimination of exchange controls, the capital movement liberalization, the enactment of the Second Banking European Directive of the European Union and the supervisory arrangements have been related to the competitive conditions of the Greek banking system.

Taking into consideration the bank assets, Spathis *et al.* (2002) distinguished banks into small and large ones. In order to investigate the differences of profitability and efficiency between small and large Greek banks, as well as the factors of profitability and operation related with the size of banks, for the years 1990-1999, a multicriteria methodology has been used.

2.2 The Efficiency of the Banking System

The efficiency of the banking system has been one of the major issues in the new monetary and financial environment. The efficiency and the competitiveness of financial institutions cannot be easily measured since their products and services are of intangible nature. Many researchers have attempted to measure productivity and efficiency of the banking industry using outputs, costs, efficiency and performance.

The scale and scope economies of banking have been one of the issues related to the competitiveness and efficiency of banks studied extensively. Murray and

White (1983), recognized the multi-product nature of financial intermediaries and used a translog cost function to evaluate the scale and scope economies of credit unions in Canada. They find that large multi-product credit unions are more cost-efficient than small single-product credit unions. Gilligan *et al.* (1984) also utilize the translog cost function to examine scale and scope economies in U.S. banking firms. They find economies of scope but not economies of scale among U.S. banks in their sample. Hunter *et al.* (1990) analyze the U.S. bank production using an intermediation approach and multi-cost production function. They find no evidence of cost complementary i.e. no sub-additive cost functions.

In addition, there are different bank performance measures that can be employed. Revell (1980) uses interest margin as a performance measure for U.S. commercial banks. He defines interest margin as difference of interest income and expense divided by total assets. Arshadi and Lawrence (1987) measure bank performance using normal correlation analysis. They consider factors related to profitability, pricing of bank services and loan market share. However, those measures of bank competitiveness are not the ones evaluated by the financial market. Size affects the efficiency of banks. Previous research, especially in the United States, indicates that scale economies appear in small banks and not in large ones (Short, 1979; Miller and Noulas, 1996). More recent research shows that the levels of size for the existence of scale economies are higher due to economic development and market liberalisation (Miller and Noulas, 1997).

It has also been proved that in this new competitive environment, large banks will survive. Small banks could only survive if they are focused on specialized activities (Peterson and Rajan, 1995; Hardy and Simigiannis, 1998). The efficiency and technical progress of German cooperative banks were examined by Lang and Welzel (1996). All banks enjoy productivity, which is higher in small banks according to this sample.

The technical efficiency of large banks was examined by Miller and Noulas (1996). Larger and more profitable banks have higher levels of technical efficiency. At the same time, larger banks are more likely to operate under decreasing returns of scale. The performance of the new US commercial banks was examined by DeYung and Hasan (1998). The profit efficiency of the new banks improves rapidly during the first years of operation, but on average it takes about nine years to reach established bank levels. Small banks lend a larger proportion of their assets to small businesses than do large banks. In the USA, Jayaratne and Wolken (1999) found that the probability of a small firm to have a line of credit from a bank does not decrease in the long run. This is done when there are few small banks in the area, although short-run disruptions may occur.

The present study uses ratios from financial statements to measure the banking performance with the aid of a multicriteria method.

3 Multicriteria Approach: The PROMETHEE Method

PROMETHEE is an outranking multicriteria decision aid method (Brans and Vincke, 1985; Brans *et al.*, 1986). The main principles upon which the PROMETHEE method is based are: a) Extension of the notion of criteria, b) development of a valued outranking relation, c) exploitation of the outranking relation.

The PROMETHEE method provides the decision-maker with a methodological framework that enables him to advance in solving a decision problem where several, often conflicting multiple criteria must be taken into consideration. The PROMETHEE II method is known to be one of the most efficient and simplest multicriteria methodologies. It is based on the outranking relations concept which was found and developed by Roy (1968). Roy defined the outranking relation as a binary relation **S** between alternatives (banks) a and b in a given set of alternatives A, such that $a\mathbf{S}b$ (a outranks b) if there are enough arguments to decide that a is at least as good as b, while there is no essential reason to refute that statement.

In this study, the construction of the outranking relation through the PROMETHEE II method involves the consideration of the performance of the banks on a set of the evaluation criteria g_1, g_2, …, g_n (financial ratios). Each criterion is assigned a weight p depending on its importance. The weight increases with the importance of the criterion. The criteria's weights constitute the basis for the assessment of the degree of preference for bank a over bank b. This degree is represented in the preference index $\pi(a, b)$:

$$\pi(a,b) = \sum_{i=1}^{n} p_j H_j(d) \Big/ \sum_{j=1}^{n} p_j$$

The preference index for each pair of banks (a, b) ranges between 0 and 1. The higher it is (closer to 1) the higher is the strength of the preference for bank a over bank b. $H_i(d)$ is an increasing function of the difference d between the performances of banks a and b on ratio g_i. $H_i(d)$ is a kind of preference intensity function (Vincke, 1992). The H_i function can have different forms, depending upon the judgment policy of the decision maker. Generally, six forms (generalized criteria) of the H function are commonly used presented in Table 1.

For the purposes of this study the Gaussian form of the H_j was used for all financial ratios. The use of the Gaussian form requires only the specification of the parameter σ. This function is a generalization of all the other five forms, whereas the fact that it does not have discontinuities contributes to the stability and the robustness of the obtained results (Brans et al., 1986).

The results of the comparisons made for all pairs of banks (a, b) are organized in a graph (value outranking graph). The nodes of the graph represent the banks under consideration, whereas the arcs between nodes a and b represent the preference of bank a over bank b (if the direction of the arc is $a \rightarrow b$) or the opposite (if the direction of the arc is $a \leftarrow b$). Each arc is associated with a flow representing the preference index $\pi(\alpha, \beta)$. The sum of all flows leaving a node a is called the leaving flow of the node, denoted by $\varphi^{+}(\alpha)$. The leaving flow provides a measure of the outranking character of bank a over all the other banks in the sample. In a similar way, the sum of all flows entering a node a is called the entering flow of the node, denoted by $\varphi^{-}(\alpha)$. The entering flow measures the outranked character of bank a compared to all the other banks. The difference between the leaving and the entering flow $\varphi(\alpha)=\varphi^{+}(\alpha)-\varphi^{-}(\alpha)$ provides the net flow for the node (bank) a which constitutes the overall evaluation measure of the performance of bank a. On the basis of their net flows the banks are ranked from the best (banks with high positive net flows) to the worst ones (banks with low net flows).

Table 1: Forms (generalized criteria) of the preference function (Source: Brans *et al.*, 1986)

Generalized criteria	Graphical representation	Parameters
I. Usual criterion	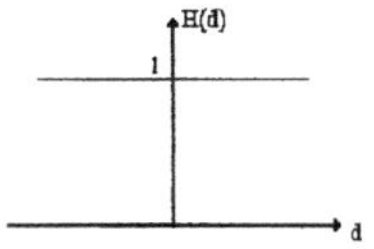	-
II. Quasi-criterion	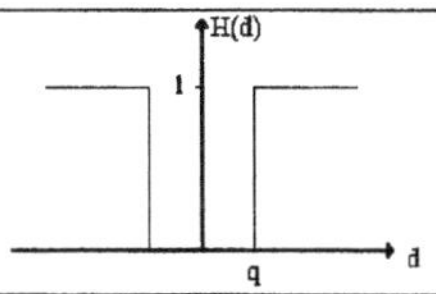	q
III. Criterion with linear preference	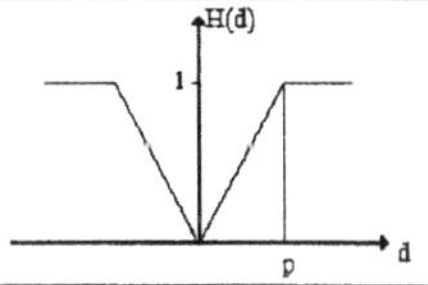	p
IV. Level criterion	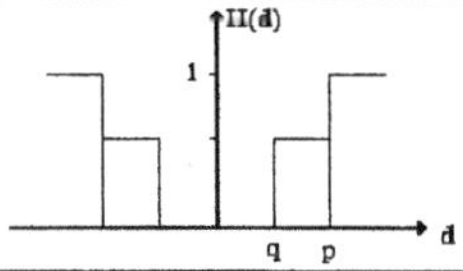	q, p
V. Criterion with linear preference & indifference area	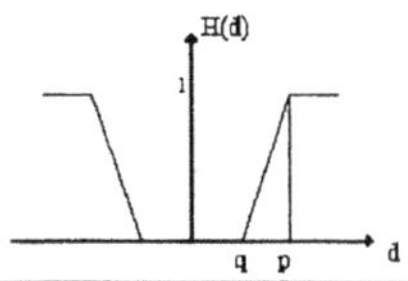	q, p
VI. Gaussian criterion	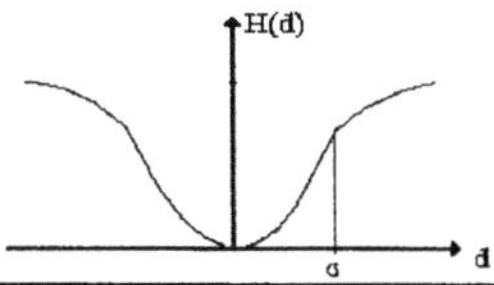	σ

4 Application

4.1 Sample and Data

The application of the above methodology was realized in a sample of Greek banks. The annual data from the financial statements of each bank separately for the period 1995-1999 were selected. The sample includes 20 Greek banks, including the Agricultural Bank of Greece, Alpha Credit Bank, Aspis Bank, Bank of Attica, General Bank of Greece, Egnatia Bank, National Bank of Greece, Commercial Bank of Greece, European Investment Bank, Ionian Bank, Credit Lyonnais Greece[1], Dorian Bank[2], Ergobank, ETEBA, ETBA, European and Popular Bank, Macedonian-Thrace Bank, Eurobank, Xiosbank and Cyprus Bank. It should be noted that the number of banks was reduced from 25 to 20 after the elimination of those for which accurate information was not completely available. Moreover, mergers and acquisitions took place among few of them.

4.2 Variables

The variables in this study involve indices based on the financial statements of banks, expressed in the form of financial ratios. The selected ratios used to evaluate the financial performance of the banks are presented in Table 2. The 15 selected financial ratios indicate the financial situation and performance of each of the bank institutions under study, in the areas of liquidity, profitability, management efficiency and capital structure, investment activity, and development.

The liquidity ratios comprise a set of measures designed to portray a firm's ability to cover its maturing short-term liabilities. According to the most widely used financial methods in banking analysis, the financial ratios cash/total deposits and quick assets/total deposits were selected to indicate liquidity capabilities of a commercial bank. Both these ratios indicate the cash position of the bank and its capacity to comply with unusual withdrawals from the depositors in emergency cases. Theoretically, the higher the ratio, the higher the depositors security. However, high liquidity could also mean improper management of resources and loss of investment opportunities. As far as the ratio quick assets/total deposits is concerned, contrary to the previous ratio where the cash is kept idle and is used only to pay short-term obligations, quick assets are funds that are invested and a certain return is expected. Some assets are closer to cash than other. In case the

[1] Credit Lyonnais Greece is currently owned by the Piraeus Bank under the name Piraeus Prime Bank.

[2] Dorian Bank is currently owned by Telesis Bank under the name Telesis Investment Bank.

cash are not enough to satisfy depositors demands, managers will turn to solutions such as securities and stocks.

Table 2: List of financial ratios

Financial ratios		Category
F1	Cash/Total deposits	Liquidity
F2	Quick assets /Total deposits	
F3	Return on assets (ROA)	Profitability
F4	Return on equity (ROE)	
F5	Net income before taxes/(Loans + Securities)	
F6	Total assets/Employees	
F7	Capital funds/Total assets	Capital Structure
F8	Securities/Capital funds	
F9	Stockholders equity/Total assets	
F10	Capital funds/Fixed assets	
F11	Securities/Total deposits	Investment Activity
F12	Loans/Total deposits	
F13	Annual change of total assets	Development
F14	Annual change of total loans	
F15	Annual change of total deposits	

Financial analysts employ the profitability ratios to judge how efficiently companies use their assets. The three ratios ROE, ROA and net income before taxes/(loans+securities) are indicative of the profitability and the earning quality of a commercial bank. The ROE ratio is the net profit after tax divided by the shareholders' equity and represents the earning performance of the bank based on the shareholder's stake. The ROA ratio is the net profit divided by total assets and represents the earning performance of the bank based on the total assets. The granting of loans is considered to be one of the most profitable services provided by commercial banks, while the investment of securities is considered to be one of the most secure services. An average value of the ratio earnings before interest and taxes/(loans+securities), that is relatively close to the average value of the ROA ratio reveals a bank strategy which is more focused on profits coming from loans and securities than from other sources of assets.

Two financial ratios are employed to analyze the investment activity of a bank. These include the loans/total deposits and the securities/total deposits ratios. The first ratio shows the extent that the bank has lent its deposits, while the second one shows the proportion of total deposits refering to securities investment.

The financial ratios equity/total assets and capital funds/fixed assets are employed to express the capital structure of a bank. According to European Union banking regulations, the value of the ratio equity/total assets must be at least 8%. Values below 8% are less desirable, while values equal to or higher than 8% are more desirable. A small value of this ratio indicates that a bank depends more on long and short-term liabilities and consumer deposits than on stockholders funds. The ratio capital funds/fixed assets indicates the financing of fixed assets by capital funds. A small value of this ratio, less than one, could mean that the institution does not have long-term resources to finance long-term investment needs.

An analysis of the financial performance of the commercial banks, in a situation of increasing concentration, mergers, acquisitions, high competition which took place after deregulation and the further integration of the Greek financial market with that of the European Union, gives this set of ratios a special importance.

The ratio annual change of total assets shows the percentage change in the assets of a commercial bank from one year to the next. Analyzing the values of this ratio for the period 1996 to 1999 shows that although the inflation ratio decreased, the average value of this ratio steadily increased. The increase in the average values in these years can be attributed not only to the banks' policy which become more favorable for investments when the Greek government began to intervene less in the Greek financial market but also and when mergers and acquisitions took place among various banks as a consequence of this free financial market (e.g., acquisition of Ioniki Bank by Alpha Bank, merger between Eurobank and Ergasias Bank, etc.).

Consumer deposits are the most important source of funds used by commercial banks to finance their activities. A high value of the ratio annual change total deposits means that the bank has sufficient funds to finance its activities, otherwise it has to shrink its activities or to find sources by borrowing.

4.3 Results

The first step of the analysis regarding the financial performance of the Greek banks was to perform a multivariate statistical analysis, namely principal component analysis (PCA). The purpose of using PCA in this study was to reduce the available amount of information so as to retain only the one that best describes the characteristics of the banks in the sample. In that regard, PCA enables the analysis of the sample along the considered financial ratios in order to measure the contribution of each ratio in explaining the variance in the sample. In this way, it is possible to form a reduced set of ratios that embodies the most relevant information for the analysis, while disregarding ratios with small explanatory power. This is a necessary stage in the analysis considering the large number of ratios (15 overall) as opposed to the small number of banks in the sample (20 banks). Furthermore, this stage has also practical implications. Reducing the number of ratios upon which to base the analysis, implies that during his daily practice an analyst is required to have available only a small amount of information in order to proceed with the evaluation of the financial performance of the banks. Therefore, a significant amount of cost and time required to gather and analyse additional irrelevant information is saved during the analysis.

PCA is applied on the averages of the financial ratios for the period under consideration (1995-1999). This specification enables the analysis of the characteristics of the banks throughout the considered period. The application of

PCA led to the construction of seven principal components with eigenvalues higher than 1. Each of these principal components is expressed as a linear combination of the 15 financial ratios used in the analysis. The loadings of each ratio in each of the seven principal components was used in order to select the most relevant ratios for the analysis. The ratios selected through this approach are presented in Table 3.

Table 3: Financial ratios selected by PCA

Financial ratios		Category
F2	Quick assets/Total deposits	Liquidity
F3	Return on assets (ROA)	Profitability
F4	Return on equity (ROE)	
F10	Capital funds/Fixed assets	Capital Structure
F12	Loans/Total deposits	Investment Activity
F13	Annual change of total assets	Development
F15	Annual change of total deposits	

Each of these seven ratios is considered as a criterion for the evaluation of the financial performance of the banks using the PROMETHEE method. However, in order to employ the PROMETHEE method the significance (weight) of each ratio needs to be specified. The weight is a very important parameter in the application of the PROMETHEE method. The decision as to what weights will be used is the exclusive responsibility of the decision-maker. If the decision-maker considers that a specific financial ratio is relatively more important than the others, then the weight coefficient of this ratio will be higher than that of the others, and vice-versa. In this way, a bank that has high performance on this ratio will have an advantage compared with the others which do not have such high performance. Nevertheless, in this study such a decision-maker (financial analyst) was not available to specify the weights of the financial ratios. To overcome this difficulty in the application of the PROMETHE method, some representative combinations of weights are considered. Table 4 presents the considered combinations of the weight coefficients for the five groups of financial ratios.

These scenarios are specified in such a way that they give neither a strong disadvantage nor a strong advantage to any of the ratios. An exception is only the sixth scenario where all groups of financial ratios have the same weight coefficient (3). Conceptually, in this scenario all dimensions of bank financial performance (groups of financial ratios) are considered of equal importance. It should also be pointed out that ratios belonging into the same group are always treated as of equal importance. This is the case of the ratios F3 and F4 which both belong into the profitability groups and the ratios F13 and F15 both belonging into the development group.

The PROMETHEE method is applied for each of the above weight scenarios. In using this method the preference function $p_i(a, b)$ is specified using the Gaussian function for all financial ratios $g_i \in$ {F2, F3, F4, F10, F12, F13, F15} and for all pairs of banks a and b: $H_i(d) = 1 - \exp(-d^2/2\sigma^2)$, where σ is the standard deviation of the ratio g_i.

Table 4: Scenarios for weight coefficients of the financial ratios

Financial ratios	Scenarios					
	1	2	3	4	5	6
Liquidity	*3.50*	*1.50*	*2.00*	*2.50*	*3.00*	*3.00*
F2	3.50	1.50	2.00	2.50	3.00	3.00
Profitability	*2.50*	*3.00*	*3.50*	*1.50*	*2.00*	*3.00*
F3	1.25	1.50	1.75	0.75	1.00	1.50
F4	1.25	1.50	1.75	0.75	1.00	1.50
Capital structure	*2.00*	*2.50*	*3.00*	*3.50*	*1.50*	*3.00*
F10	2.00	2.50	3.00	3.50	1.50	3.00
Investment activity	*3.00*	*3.50*	*1.50*	*2.00*	*2.50*	*3.00*
F12	3.00	3.50	1.50	2.00	2.50	3.00
Development	*1.50*	*2.00*	*2.50*	*3.00*	*3.50*	*3.00*
F13	0.75	1.00	1.25	1.50	1.75	1.50
F15	0.75	1.00	1.25	1.50	1.75	1.50

Table 5 presents ranking of the banks according to their net flows which are obtained from the application of the PROMETHEE method for each of the weight scenarios. According to these results the ETEVA Bank is in the first position of the rankings for all the scenarios. This bank has the best results in ratios F3 and F15. It should also be mentioned that although the ETEVA Bank has very good results in almost all ratios, the profitability ratios are the criteria that make the difference for this bank compared with the others (almost in every year ETEVA is the most profitable bank). ETEVA is followed by Credit Lyonnais Greece (now owned by Piraeus Bank, under the name Piraeus Prime Bank), Eurobank and the ETVA Bank. With regard to the ETVA Bank it should be noted that its evaluation is quite high in all weight scenarios, except scenario 3. In this scenario more significance is given to the profitability and investment ratios, where the bank's performance is rather low. This justifies its low position in the ranking for this specific scenario.

On the other hand, the European and Popular Bank, the Dorian Bank (Telesis Investment Bank), Xiosbank and the National Bank of Greece appear in the last positions of the rankings. The evaluation of these banks is low for all scenarios. This is indicated by the low positions that these banks have in the rankings obtained from the application of the PROMETHEE method for each scenario.

Overall, it should be noticed that the differences between the results obtained for the different weight scenarios are quite limited. This is indicated by the value of the Kendall's coefficient of concordance *W*, which is 0.909 (significant at the 1% level).

Table 5: Ranking of the banks according to the PROMETHEE method

Banks	Scenarios						Average rank
	1	2	3	4	5	6	
Alpha Credit Bank	4	6	5	7	6	6	5.7
Agricultural Bank of Greece	10	12	12	17	16	13	13.3
Aspis Bank	17	16	17	9	8	15	13.7
Bank of Attica	16	17	16	11	14	16	15.0
Commercial Bank of Greece	11	9	8	13	15	10	11.0
Credit Lyonnais Greece	3	2	2	2	3	2	2.3
Cyprus Bank	14	10	9	6	10	9	9.7
Dorian Bank	19	19	19	16	17	19	18.2
Egnatia Bank	8	8	7	10	9	8	8.3
European Investment Bank	6	5	4	5	5	5	5.0
European and Popular Bank	20	20	20	20	20	20	20.0
ETVA Bank	2	3	11	4	2	3	4.2
ETEVA Bank	1	1	1	1	1	1	1.0
General Bank of Greece	12	11	10	12	12	11	11.3
Ionian Bank	9	15	14	15	11	12	12.7
Macedonian-Thrace Bank	15	13	13	14	13	14	13.7
National Bank of Greece	13	14	15	19	18	17	16.0
Eurobank	5	4	3	3	4	4	3.8
Ergobank	7	7	6	8	7	7	7.0
Xiosbank	18	18	18	18	19	18	18.2

5 Conclusions

The present study evaluated the financial performance of 20 commercial banks operating in Greece using the multicriteria method PROMETHEE. The banks in this study were evaluated according to the following five groups of ratios: liquidity, profitability, investment activity, capital structure and development.

PCA revealed the most important ratios which were applied to the multicriteria method PROMETHEE as criteria in order to evaluate the financial performance of the Greek commercial banks. The results obtained from the multicriteria approach for different weight scenarios were quite robust and led to helpful conclusions on the competitive financial performance of the banks.

This kind of analysis can be quite helpful to bank managers, investors, portfolio analysts and policy makers. Banks are a crucial point in the development of the financial and business environment of every country. Therefore, it is of major importance to develop an in-depth evaluation and monitoring framework regarding banking performance. The methodology presented in this study contributes towards this direction. Of course, banking performance is not solely based on the financial characteristics of the banks. Other issues also need to be considered (management, organization, market niche, etc.). Nevertheless, the vulnerable financial environment worldwide, has increased the interest towards the development and implementation of sound methodological tools for financial analysis and launching in the banking sector.

References

Alexakis, P., Thomadakis S. and Xanthakis, M. (1995), "Banks liberalization and profitability: Evidence from Greek commercial banks", *Journal of International Financial Markets, Institutions and Money*, 5/2-3, 181-192.

Arshadi, N. and Lawrence, E.C. (1987), "An empirical investigation of new bank performance", *Journal of Banking & Finance*, 11, 33-48.

Brans, J.P. and Vicke, P. (1985), "A preference ranking organization method", *Management Science*, 31, 647-655.

Brans, J.P., Vincke, Ph. and Mareschal, B. (1986), "How to rank and how to select projects: The PROMETHEE method," *European Journal of Operational Research*, 24, 228-238.

DeYung, and Hasan, I. (1998), "The performance of de novo commercial banks: A profit efficiency approach", *Journal of Banking & Finance*, 22/5, 565-587.

Garganas, N. (1998), "Forecasting the impact from the introduction of the EURO to the financial markets and the banking systems", *Journal of the Hellenic Bank Association*, 16/5-8, 19.

Gilligan, T., Smirlock, M. and Marshall, W. (1984), "Scale and scope economics in the multi-product banking firm", *Journal of Monetary Economics*, 13, 393-405.

Gortsos, C. (1998), *The Greek Banking System*, Hellenic Bank Association, Athens, Greece (in Greek).

Hardouvelis, G. (1997), "Financial markets and banks to the Euro zone in the present and in the future of the Greek economy", A. Kintis (Ed.), vol. A, Economic University of Athens, Gutenberg Publishers, 684-703 (in Greek).

Hardy, C.D. and Simigiannis, G.T. (1998), "Competitiveness and effectiveness of the Greek banking system", *Economic Review of the Bank of Greece*, 11, 7-30 (in Greek).

Hondrogiannis, G., Liolios, S. and Papapetrou, E. (1999), "Assessing competitive conditions in the Greek banking system", *Journal of International Financial Markets, Institutions and Money*, 9/4, 377-391.

Hunter, W.C., Timme, S.G. and Yang, W.K. (1990), "An examination of cost subadditivity and multiproduct production in large U.S. commercial banks", *Journal of Money, Credit and Banking,* 22, 504-525.

Jayaratne, J. and Wolken, J. (1999), "How important are small banks to small business lending? New evidence from a survey of small firms", *Journal of Banking & Finance,* 23, 427-458.

Karafolas, S. and Mantakas, G. (1996), "A note on cost structure and economies of scale in Greek banking", *Journal of Banking & Finance*, 20/2, 377-387.

Karamouzis, N. (1998), "The transition of the Greek banking system to the Euro zone", *Journal of the Hellenic Bank Association,* 13, 33-42 (in Greek).

Kloutsiniotis, A. (1996), "The European commercial banks in the framework of qualitative changes to the national banking systems" *Economic Review of Commercial Bank of Greece,* 8, 4-19 (in Greek).

Kosmidou, K. and Spathis, Ch. (2000), "Euro and profitability of Greek banks", *European Research Studies Journal* (in press).

Lang, G. and Welzel, P. (1996), "Efficiency and technical progress in banking empirical results for a panel of German cooperative banks", *Journal of Banking & Finance,* 20/ 6, 1003-1023.

Miller, S. and Noulas, A. (1996), "The technical efficiency of large banks production", *Journal of Banking & Finance*, 20/3, 495-509.

Miller, S. and Noulas, A. (1997), "Portfolio mix and large-bank profitability in the USA", *Applied Economics*, 29, 505-512.

Murray, J.D. and White, R.W. (1983), "Economies of scales and economics of scope in multiproduct financial institutions: A study of British Columbia credit unions", *Journal of Finance*, 38, 887-902.

Noulas, A. (1999), "Profitability and efficiency of Greek banks (1993-1998)", *Journal of the Hellenic Bank Association*, 19/20, 53-60 (in Greek).

Papaioannou, M. and Ganzonas, E.K. (1997), *Financial Innovation Involving the Greek Drachma.* IMF Working Paper 97/14 (Washington: International Monetary Fund).

Peterson, M. and Rajan, R. (1995), "The effects of credit market competition on lending relationships", *Quarterly Journal of Economics*, 60, 407-444.

Revell, J. (1980), "Costs and margins in banking: An international survey", Paris, OECD.

Roy, B. (1968), "Classement et choix en présence de points de vue multiples: La méthode ELECTRE", *R.I.R.O*, 8, 57-75.

Short, B. (1979), "The relation between commercial bank profit rates and banking concentration in Canada, Western Europe and Japan", *Journal of Banking & Finance,* 3, 209-219.

Spathis, Ch. and Kosmidou K. (1999), "Competitiveness of small & large Greek banks: An investigation", in: C. Agorastos (ed.), *Proceedings of the International Conference "Preparing the Manager of the 21st Century"* (vol. I), 205-218.

Spathis, Ch., Kosmidou, K. and Doumpos, M. (2002), "Assessing profitability factors in the Greek banking system: A multicriteria methodology", *International Transactions of Operational Research* (to appear).

Staikouras, C. and Steliaros, M. (1999), "Determinants factors of profitability of the Greek banking system", *Journal of the Hellenic Bank Association*, 19/20, 61-66 (in Greek).

Tolios, G. (1998), "Banking and insurance groups in the Greek economy. Basic characteristics and development trends", *Economic Review of the Commercial Bank of Greece,* 16, 13-25 (in Greek).

Vasiliou, D. (1992), "Which are the most profitable banks in Greece, the small, the medium or the big ones?". *Hellenic Bank Association*, 73-80 (in Greek).

Vasiliou, D. (1993), "Investigation of profitability of Greek bank market", *Journal of the Hellenic Bank Association,* 37-38 (in Greek).

Vincke, P. (1992), *Multicriteria Decision Aid,* John Wiley & Sons Ltd, New York.

Zopounidis, C., Despotis, D.K. and Stavropoulou, E. (1995), "Multiattribute evaluation of Greek banking performance", *Applied Stochastic Models and Data Analysis*, 11/1, 97-107.

Zopounidis, C., Doumpos, M. and Georgiou, H. (1997), "Application of a multicriteria methodology in the assessment of banking branches efficiency", *Economic Review of the Commercial Bank of Greece,* 10, 4-9 (in Greek).

Business Failure and Credit Risk Assessment

Business Failure Prediction: A Review and Analysis of the Literature

M. Daubie, N. Meskens

Catholic University of Mons (Fucam),
Production and Operations Management Department
151 Chaussée de Binche
B-7000 Mons, Belgium

Summary:
Business failure prediction is a topic of great importance for a lot of people (shareholders, banks, investors, suppliers,...). That's why a lot of models were developed in order to predict it. Statistical procedures (multiple discriminant analysis, logit or probit) were among the most used methods in this kind of problem. However, parametric statistical methods require the data to have a specific distribution. In addition to the restriction on the distribution involved, multi-collinearity, autocorrelation and heteroscedasticity could lead to problems with the estimated model with some statistical methods. Because of these drawbacks, others methods have been investigated : multicriteria methods (i.e. UTA, Electre tri,...) or machine learning methods (i.e. neural network, genetic algorithm, decision tree, instance based learning,...). Our main target is to provide a review of the literature but also to have a larger view than usually by evoking causes, symptoms and remedies of bankruptcy. We also proposed new perspectives and topics of research.

Keywords: Business failure prediction, Review of literature

1 Introduction

Since Altman's seminal work (Altman [1968]) until nowadays, the business failure prediction problem received a lot of interest from both researchers and academicians. This interest is reflected by the large number of papers published since 1968. This problem is of great importance for two principal reasons. Firstly, the number of bankruptcies increased dramatically all over the world. For example, in Belgium, there were 7150 bankruptcies in 1999 against 3907 in 1990. Secondly, people involved in a business failure are large and various: workers, owners or shareholders, managers, lenders, suppliers, clients, the community, the government,.... As the number of people involved in bankruptcy are numerous, the cost of failure is important.

Failure can be defined by different ways but is often expressed as the inability of the firm to pay its financial obligations when they come due. However, because of practical use, authors often choose company which filed a formal legal document in court for the purpose of either liquidation or reorganization. Some authors chose different kind of dependent variables like the audit qualification given by auditors or accountants (Lacher et al. [1995]) or loan default (Abdel-Khalik and El-Sheshai [1980] and Chalos [1985]). We must also mention that bankruptcies is not the only "possibility" for firms encountering difficulties. Reorganization (Casey et al. [1986]) or merger of the distressed firm (Zanakis and Zopounidis [1997]) are other solutions.

Some authors tried to summarize the considerable efforts made in this field. Vernimmen [1978] reviews some failure prediction models and gives their contributions and limits. Scott [1981] compares the major types of empirical and theoretical models in order to identify the overlap between them. Altman [1984] surveys some models developed outside US for failure prediction and Keasy and Watson [1991] assessed the uses and limitations of different models. Zopounidis and Dimitras [1993] present the main methods employed for prediction of business failure with their advantages and disadvantages. Dimitras et al. [1996] provide an extensive review of the literature based on a framework which includes classification of the studies by countries, method and industrial sector. Guilhot [2000] recalls the evolution of different approaches (economic, financial, strategic and managerial) used in business failure prediction.

The aim of this paper is to provide an up-to-date review of the literature of this important topic. Our review is useful and original for two main reasons. Firstly, our view of the bankruptcy literature is larger than usual. Indeed, we are deeply interested not only in the methods and variables (symptoms of failure) used in studies but also in causes, process and remedies of failure. Secondly, in our survey of methods, even if we focus mainly on last developments occurred during the last five years, we adopt a chronological point of view, from Beaver [1966] to nowadays. It allows us to propose some new perspectives to tackle this problem.

The remainder of the paper will be organized as follows. In section 2, we spend some time to present causes, process and remedies of business failure. Following that we introduce in section 3 symptoms used to reveal the failure. Before the final section containing concluding remarks, a survey of methods for bankruptcy prediction applications is presented.

2 Causes, Process and Remedies

" The failure process of a firm can be compared with the evolution of a disease in a human being. In the same way, failure itself is comparable to death. The causes of failure are often associated with management adequacy. These cause lead to occurrence of symptoms which are observable from the deterioration of financial ratios. This deterioration factually forms the failure process". From this statement given by Luoma and Laitinen [1991], we can infer that (a) failure is not a sudden event, it takes some time ("evolution of the disease") for a company to become bankrupt (b) financial ratios are only symptoms of failure and (c) failure process is triggered off by something (causes). We could also add to this statement that as in human being, if disease is revealed early enough, some remedies can be taken. Unfortunately, very few authors are interested in causes, process or remedies. They work only with symptoms of failure, often represented by financial ratios.

As mentioned earlier, very few authors (Meyer and Pifer [1970], Altman [1971], Vernimmen [1978], Dambolena and Khoury [1980], Sharma and Mahajan [1980], Koenig [1985] or Makridakis [1991]) are really interested in origins of bankruptcy. Even if the list of causes may be different across studies, we can distinguish two main causes of failure: endogenous and exogenous factors. The main cause within organisation is managerial incompetence. This is the case in France, Belgium but also in US where 90% of bankruptcies can be impute to this factor (Sharma and Mahajan [1980]). Dambolena and Khoury [1980] add that bad management appears through lack of responsiveness to change in technology, insufficient consideration for cost factors, poor knowledge of financial matters,.... Meyer and Pifer [1970] and Makridakis [1991] corroborate the others but add also the integrity of employees, the organizational arteriosclerosis, the undervaluation of the competition or the believe in barriers to entry as factors explaining failures.

What exogenous factors concern, Altman [1971] mentions that the change in the nation's failure rate is negatively associated with changes in overall economic activity (Guilhot [2000] mentions that there is a systematic correlation between failures and economic crisis), stock market performance, and money supply conditions. That's why some authors (Altman [1971], Mensah [1984],...) include in their model macroeconomic factors. The influence of macroeconomic factors on bankruptcy seems particularly important because, according to Graydon (a leading service providers in the field of Credit Management and database business information), the number of bankruptcies in 2000 is decreasing in Belgium thanks to the relatively good and stable economic situation in our country. Moreover, Vernimmen [1978] signals that the failure of one firm's customer was the main cause of firm's bankruptcy in France. This last statement reinforces what we said in the introduction: bankruptcy prediction is a very important topic because failure of a particular company could have unfortunate impact on other entities.

As emphasised by Luoma and Laitinen [1991], failure is not a sudden event, it follows a kind of "path". Some authors (Ooghe and VanWymeersch [1986], Luoma and Laitinen [1991] or Laitinen [1992]) looked into the "failure path". Even if the process can be very different across distressed companies, it seems that we can identify some constants. The starting point of the process is a high interest charges burden mixed with an insufficiency of revenue and a poor profitability. When the availability of share capital is limited, firms are forced to get more indebted (first with long terms and next with short terms). This is the beginning of a vicious circle that ends in bankruptcy. This increasing indebtedness leads to a poor solvency, creditors becomes suspicious and liquidity problems arises. Thus final symptoms come to light through an increase in short-term debt which leads to poor liquidity and long average payment period for accounts payable. According to Luoma and Laitinen [1991], in the same time the firm may use "creative accounting" to show better financial figures.

As we can see, the failure process is dynamic, ratios useful to detect financial distress in the beginning of these problems are not the same as these at the end of the process. Through multidimensional scaling, Mar Molinero and Ezzamel [1991] has also point out this dynamic nature of the financial problems.

The majority of the papers considered here are interested in the prediction of bankruptcy but not in avoidance of these serious problems. Nevertheless, it would be much more fruitful for companies in difficulties to know how to avoid the failure. A help system to distressed firm must be articulated in 3 stages: a diagnosis, a recovery plan and the following up of the plan fulfilment. Even if Laitinen [1992] gives some very general preventive advice to entrepreneurs and managers of newly founded companies, these advice are general and unrelated with a particular diagnosis.

On the other hand, the expert system of Coleman et al. [1991] can recommend remedial action to improve the financial condition of the company. This is a vital capability as the system is not only able to pinpoint potential problems but also can suggest ways to avoid problems. Moreover, this system is able to play "what-if" games to determine the efficiency of remedial action. In the same way, Lincoln [1984] not only develops models to measure the levels of insolvency risk for firms in different industries but he can also suggest some remedies and calculate the impact of alternative courses of action on its insolvency risk.

3 The Symptoms of Failure

In section 2 we have already point out that most of authors rely on symptoms of the disease in order to predict failure or bankruptcy. These symptoms can be qualitative or quantitative (financial ratios) variables.

In fact, a large number of ratios has been proposed in the literature. We have tried to count the ratios used in all applications at our disposal since the work of Altman [1968] (excluding applications on too typical sectors like banking or insurance). More than 250 different (even slightly) ratios were used. It's strictly impossible to give here all these ratios. We have listed, in Appendix C, ratios used in at least 7 applications published in the last five years (see Table 1, Section 5). We have excluded from the appendix C the applications on too typical sectors (like banking or insurance). In appendix C, the number in the first colum refers to the number of the paper in Table 1. Some authors are not present in Appendix C because they don't mention the ratios they used or because they use very peculiar ratios. From appendix C, we notice that the most frequently used financial ratios are CA/CL (used in 13 applications on 21), WC/TA, EBIT/TA, QA/CL and NI/TA with 12, 11, 11 and 10 occurrences respectively. It is not surprising since they were already pointed out by Dimitras et al. [1996]] who give also an extensive list of financial ratios. It is remarkable to note that the 5 ratios included in the Altman [1968]'s model are all in the top of this table.

Rapidly after his initial work, Altman was aware that it was necessary to include non-financial and more qualitative information in the model (Altman and Loris [1976]). Others authors were agree with him (for example Collongues [1977] and Vernimmen [1978]). However, only 5 papers (Cormier et al. [1994], Dimitras et al. [1995], Slowinski and Zopounidis [1995], Greco et al. [1998] and Zopounidis and Doumpos [1999b]) really include qualitative information in bankruptcy prediction model. The last 4 papers include as qualitative information managers' work experience, firm's market niche-position, special competitive advantage of firms or market flexibility. Cormier et al. [1994] include other qualitative indicators like investment in a new sector, change in the depreciation method or change in the ownership.

Few authors (Abdel-Khalik and El-Sheshai [1980] or Chalos [1985]) use trend data in the form of the average change over several years in different ratios. It seems that trend variables have a greater ability to draw out information from balance sheet data and more able to depict "creative accounting".

Other important variables were also put forward by researchers. For example, industry type seems to be of critical importance in failure prediction and therefore was included in some models. Some authors also advise to develop a separate model for each industry (Mader [1975], Mensah [1984], Taffler [1984], Izan [1984] or Chalos [1985]) even if Bhargava et al. [1998] proved the contrary. The size of the firm seems also have an impact on the financial distress of a firm. Consequently, authors incorporate the size represented by, for example, the total assets, sales (Jo and Han [1997]), the total number of employees (Lennox [1999]), the ratio Total Assets / GNP price level (Bryant [1997]) or ln(net sales) (Laitinen [1992]).

4 Survey of Methods

After an extensive but not necessarily exhaustive review of journal articles written in English or French, we identified a list of 150 references (available upon request by the authors) on failure prediction. These papers were published in the period 1968-2000 in various journals specialized in accounting, finance, operations research and decision science. Readers should be cautious in interpreting the results of this survey since the findings are based on data collected only from journal articles. Conferences proceedings, working papers and doctoral dissertations were excluded except if they were directly and easily at our disposal.

Our approach of the methods will be twofold. First of all, we will have a chronological point of view. This kind of approach allow us to glance at the history of business failure prediction methods. Secondly, we will analyse the most recent papers on business failure prediction with a classical framework, namely Authors, Year of publication, Country and Methods.

If we start from the second War World, the first researcher to work on business failure prediction was Beaver [1966] who made the most contributive Univariate analysis of the problem. He was followed by the famous researcher on this topic i.e. Altman. Altman [1968] performed a Multivariate analysis of failure by means of multiple discriminant analysis. The discriminant analysis was rapidly extended to different sectors (bank failure prediction by Meyer and Pifer [1970] or railroad bankruptcy by Altman [1971]). In spite of discriminant analysis' drawbacks (it's beyond the scope of this paper to assess advantages and disadvantages of each method and how to overcome their limitations. People interested in these issues may see Dimitras et al. [1996] or Zanakis and Walter [1994]) it takes about 10 years before the first application of Logit model on this topic (Martin [1977]). The beginning of '80 were marked by numerous works using Human expert in order to detect failure (Zimmer [1980] or Abdel-Khalik and El-Sheshai [1980]). In 1984 we see the first Probit models applied to corporate distress (Izan [1984] and Zmijewski [1984]). During the same year, Takahashi et al. [1984] used Principal Component Analysis for bankruptcy prediction.

The second part of the eighties was a true turning in bankruptcy prediction. Indeed, between 1985 and 1988 researchers leave the statistical sphere and enter in the Multicriteria era (Zopounidis [1985]) or in the Machine Learning world (Frydman et al. [1985]). In 1988, Elmer and Borowski [1988] used the first Expert System for the financial analysis of Savings & Loans. The first application on insurance institution was the Ambrose and Seward [1988]'s paper. 1990 was also the beginning of a new and prosperous epoch in failure prediction: the first Neural Network approach was born (Tam and Kiang [1990]). It was the beginning of a very long series. Tam and Kiang [1990] were also the first to apply K Nearest Neighbour to failure prediction. Contrary to Neural Networks, Survival Analysis and Data Envelopment Analysis didn't encounter a lot of success (Luoma and Laitinen [1991] and Fernandez-Castro and Smith [1994]). Neural Networks were not only often used in the classical way (supervised learning algorithm) but they were also used as a clustering method i.e. as an unsupervised algorithm, the Self-Organizing Map (Martin-del-Brío and Serrano-Cinca [1993]).

Given the great number of methods used for bankruptcy prediction, it was obvious that someone decide to combine some of them. The first Hybrid method was created by Markham and Ragsdale [1995] who mixed Discriminant Analysis and Neural Networks. They were followed by Jeng et al. [1997], Gorzalczany and Piasta [1999] or Ahn et al. [2000]. It's important to notice that the two first applications were the first ones to use Fuzzy sets in business failure prediction.

From 1995 until nowadays, a lot of new methods were used: Rough Sets (Slowinski and Zopounidis [1995]), Genetic Algorithms (Back et al. [1996]), the Case Based Forecasting (Jo and Han [1996]), the Multi-Factor Analysis (Vermeulen et al. [1998]) or a Multiplicative hazards model (Honjo [2000]). Recently again, Zopounidis and Doumpos [1998] developed a Multicriteria Decision Support System which is a promising tool in financial diagnosis.

As we can see through this rapid and chronological overview, a lot of different methods were used starting from statistical methods at the beginning to the extended use of machine learning methods in the nineties.

Our chronological survey of business failure prediction methods is a good starting point but it seems obvious that emphasis should be put on more recent applications. That's why Table 1 contains a list of recent publications. In order to keep this review manageable, our analysis contains only journal articles published since 1996 and presenting an application of a method on bankruptcy prediction. Therefore we eliminate all review papers but also all papers which focus on bankruptcy prediction but without doing a true application of a particular method.

Table 1 contains 37 applications described by Author(s), Year of publication, Country and Methods used (Appendix A provides a list of abbreviations). People interested in the same kind of analysis for older applications, can see Zopounidis and Dimitras [1993] or Dimitras et al. [1996] or ask the authors for the same table since 1990.

Table 1: Characteristics of failure studies

N°	Authors (year of publication)	Countries	Methods
1	Back et al. [1996]	Finland	Hybrid (DA or LA or GA + NN)
2	Ignizio and Soltys [1996]	US	NN, DA
3	Jo and Han [1996]	Korea	DA,NN, CBF
4	Lee et al. [1996]	Korea	Hybrid (MDA or Tree or SOFM +NN)
5	Leshno and Spector [1996]	US	NN
6	Rahimian et al. [1996]	US	NN
7	Serrano-Cinca [1996]	US	SOFM
8	Tan [1996]	Australia	NN, PA
9	Bardos and Zhu [1997]	France	NN, DA, LA
10	Bell [1997]	US	NN, LA
11	Bryant [1997]	US	CBR, LA
12	Jo and Han [1997]	Korea	NN, CBR, DA
13	Jeng et al. [1997]	/	Hybrid (Tree+Fuzzy), Tree, DA
14	Olmeda and Fernandez [1997]	Spain	NN, DA, LA, tree

Table 1: Characteristics of failure studies (*continued*)

N°	Authors (year of publication)	Countries	Methods
15	Bardos [1998]	France	DA
16	Bhargava et al. [1998]	CA & US	LA
17	Greco et al. [1998]	Greece	RS
18	Kiviluoto [1998]	Finland	SOM
19	Laitinen and Latinen [1998]	Finland	Human Expert
20	McGurr and DeVaney [1998]	US	MDA
21	Varetto [1998]	Italy	GA, DA
22	Piramuthu et al. [1998]	Belgian	Hybrid (Tree + NN)
23	Vermeulen et al. [1998]	/	Multi-Factor
24	Zopounidis et al. [1998]	Greece	DA, Electre tri
25	Zopounidis and Doumpos [1998]	Greece	FINCLAS
26	Dimitras et al. [1999]	Greece	RS, DA, LA
27	Gorzalczany and Piasta [1999]	US	Hybrid (NN+fuzzy), RS, Tree
28	Kim and McLeod [1999]	US	Human Expert, DA, LA, Tree, NN
29	Lennox [1999]	UK	DA, LA, PA
30	Zhang et al. [1999]	US	NN, LA
31	Zopounidis and Doumpos [1999b]	Greece	UTADIS, DA
32	Zopounidis and Doumpos [1999a]	Greece	UTADIS, DA, PA, LA
33	Ahn et al. [2000]	Korea	Hybrid(RS+NN), NN, DA
34	Alam et al. [2000]	US	Fuzzy clustering, SOM
35	Etheridge et al. [2000]	US	NN
36	Honjo [2000]	Japan	Hazards model
37	Shah and Murtuza [2000]	US	SOM

Most of applications (number 1, 3, 12, 21, 22, 24, 25, 26, 29, 31 and 33) contained in Table 1 concern several sectors. However, some applications concern particular sectors like banking (8, 10, 14, 34 and 35), industrial (17, 18, 28 and 32), manufacturing (15, 19, 27, 30 and 36), retailing (20), both retailing and manufacturing (11 and 16) or computers (37). Since 37 applications were published between 1996 and 2000 (and 70 since 1990) we can say that bankruptcy prediction was really a big problem of the last years. As we can also see from Table 1, failure is an international problem because 12 different countries are represented in this table (Dimitras et al. [1996] achieved the same results but not with the same countries and for the period 1932-1994). Most of them comes from the United-States. As far as the methods are concerned, we notice that about 15 different methods were used to tackle the failure prediction problem. Nevertheless, it seems clear that Neural Networks, supervised or unsupervised, dominate the majority of applications. However, Discriminant Analysis is also used frequently, often as a benchmark for others methods.

5 Concluding Remarks and Further Research

The goal of this paper was to provide an up-to-date review of the literature on the business failure prediction problem. One originality of our review is that we spend time on causes, process and remedies of failure. For us, it's of prime interest to waste some time on these topics. Indeed, we believe that a better comprehension of the causes and the process of failure can lead to a better choice of the variables used to detect problems and therefore to a better model. Besides, exploiting qualitative variables as symptoms of failure has already proved to be very promising. Nevertheless, some efforts should yet be done in this direction in order to enhance models. Moreover, given that the dynamic nature of the failure process has rarely been used in failure prediction, it seems also useful to go deeply into these question. As mentioned in section 2, making proposal of remedial actions would be of greatest interest for firms in financial distress. It's now time to switch from a prediction point of view towards a prevention perspective. Business failure prediction is one step, failure avoidance is one step further.

In Section 4, we have seen that a lot of different methods were used for business failure prediction. Some authors have already proposed to combine of some of them. However, it's authors' opinion that going further into integration and/or hybridisation of different methods will be an exciting domain of further research. Keeping the same idea of mixing different methods, Data Mining would also be of great interest for our topic. In applications we surveyed, an important aspect was often neglected: the Decision Maker. It is also our opinion that in future development of new methods, he and its preferences must be central like in multicriteria methods.

Appendix

Appendix A: Abbreviations used for methods in Table 1 (Section 5)

CBF	Case Based Forecasting	NN	Neural Networks
CBR	Case Based Reasoning	NPDA	Non-Parametric DA
(L)(M)(Q)DA	(Linear)(Multiple)(Quadratic) Discriminant Analysis	RS	Rough Sets
Finclas	Financial Classification	PA	Probit Analysis
GA	Genetic Algorithm	SO(F)M	Self Organizing (Feature) Map
LA	Logit Analysis	UTADIS	DIScriminant Additive UTilities

Appendix B : Abbreviations used for ratios in Appendix C

CA	Current Assets	QA	Quick Assets
CL	Currents Liabilities	RE	Retained Earnings
EBIT	Earning Before Interest and Taxes	TA	Total Assets
MVE	Market Value of Equity	TL	Total Liabilities
NI	Net Income	WC	Working Capital
NW	Net Worth		

Appendix C: Most frequently financial ratios included in industrial or retailing applications

Study number used in Table 1	CA / CL	WC / TA	EBIT / TA	QA / CL	NI / TA	NI / NW	RE / TA	TL / TA	MVE / TD	CL / TA
1	+	+	+	+	+		+	+	+	
2		+	+				+		+	
4	+			+	+	+				
5		+	+	+			+	+	+	
6		+	+				+		+	
7		+	+				+		+	
11	+	+		+	+	+	+	+		+
12	+	+	+	+	+	+				+
17			+			+		+		
19			+	+				+		
20	+				+					+
21										
23	+	+		+				+		
25	+			+	+	+				+

Appendix C: Most frequently financial ratios included in industrial or retailing applications (*continued*)

Study number used in Table 1	CA/CL	WC/TA	EBIT/TA	QA/CL	NI/TA	NI/NW	RE/TA	TL/TA	MVE/TD	CL/TA
26	+	+		+	+	+				+
27		+	+				+		+	
30	+	+	+				+		+	
31	+		+	+	+	+		+		+
32	+			+	+	+				+
33	+	+			+					
37	+					+				
Total	13	12	11	11	10	9	8	7	7	7

References

Abdel-Khalik, R. and K. El-Sheshai (1980). "Information Choice and Utilization in an Experiment on Default Prediction." *Journal of Accounting Research* **18**(2): 325-342.

Ahn, B., S. Cho and C. Kim (2000). "The integrated methodology of rough set theory and artificial neural network for business failure prediction." *Expert Systems with Applications* **18**: 65-74.

Alam, P., D. Booth, K. Lee and T. Thordarson (2000). "The use of fuzzy clustering algorithm and self-organizing neural networks for identifying potentially failing banks: an experimental study." *Expert Systems with Applications* **18**: 186-199.

Altman, E. (1968). "Financial Ratios, Discriminant Analysis and The Prediction of Corporate Bankruptcy." *The Journal of Finance* **XXIII**(4): 589-609.

Altman, E. (1971). "Railroad bankruptcy propensity." *The Journal of Finance* **XXVI**(2): 333-345.

Altman, E. (1984). "The Success of Business Failure Prediction Models: An International Survey." *Journal of Banking and Finance* **8**: 171-198.

Altman, E. and B. Loris (1976). "A Financial Early Warning System for Over-The-Counter Broker Dealers." *The Journal of Finance* **31**(4): 1201-1217.

Ambrose, J. and J. Seward (1988). "Best's Ratings, Financial Ratios and Prior Probabilities in Insolvency Prediction." *The Journal of Risk and Insurance*: 229-244.

Back, B., T. Laitinen and K. Sere (1996). "Neural Networks and Genetic Algorithms for Bankruptcy Predictions." *Expert Systems With Applications* **11**(4): 407-413.

Bardos, M. (1998). "Detecting The Risk of Company Failure at the Banque de France." *Journal of Banking and Finance* **22**: 1405-1419.

Bardos, M. and W. Zhu (1997). "Comparaison de l'analyse discriminante linéaire et des réseaux de neurones. Application à la détection de défaillance d'entreprises." *Revue de Statistique Appliquée* **XLV**(4): 65-92.

Beaver, W. (1966). "Financial ratios as predictors of failure." *Empirical research in accounting : selected studies, supplement to Journal of Accounting Research* **4**: 71-111.

Bell, T. (1997). "Neural nets or the logit model? A comparison of each model's ability to predict commercial bank failures." *Intelligent Systems in Accounting, Finance and Management* **6**: 249-264.

Bhargava, M., C. Dubelaar and T. Scott (1998). "Predicting bankruptcy in the retail sector: an examination of the validity of key measures of performance." *Journal of Retailing and Consumers Services* **5**(2): 105-117.

Bryant, S. (1997). "A case-based reasoning approach to bankruptcy prediction modeling." *Intelligent Systems in Accounting, Finance and Management* **6**: 195-214.

Casey, C., V. McGee and C. Stickney (1986). "Discriminating Between Reorganized and Liquidated Firms in Bankruptcy." *The Accounting Review* **LXI**(2): 249-262.

Chalos, P. (1985). "Financial Distress: A comparative Study of Individual, Model and Committee Assessments." *Journal of Accounting Research* **23**(2): 527-543.

Coleman, K., T. Graettinger and W. Lawrence (1991). "Neural Networks for Bankruptcy Prediction: The Power to Solve Financial Problems." *AI Review* **July/August**: 48-50.

Collongues, Y. (1977). "Ratios financiers et prévision des faillites des petites et moyennes entreprises." *Banque* **365**: 963-970.

Cormier, D., M. Magnan and B. Morard (1994). "Evaluation de la pérennité de l'entreprise dans un contexte d'audit." *Economie & Comptabilité* **186**(Mars/Avril): 23-34.

Dambolena, I. and S. Khoury (1980). "Ratio Stability and Corporate Failure." *The Journal of Finance* **XXXV**(4): 1017-1026.

Dimitras, A., R. Slowinski, R. Susmanga and C. Zopounidis (1999). "Business failure prediction using rough sets." *European Journal of Operational Research* **114**: 263-280.

Dimitras, A., S. Zanakis and C. Zopounidis (1996). "A survey of business failures with an emphasis on prediction methods and industrial applications." *European Journal of Operational Research* **90**: 487-513.

Dimitras, A., C. Zopounidis and C. Hurson (1995). "Une méthode d'aide multicritère à la décison pour l'estimation du risque de faillite." *Foundation of Computing and Decision Sciences* **20**(2): 99-112.

Elmer, P. and D. Borowski (1988). "An Expert System Approach to Financial Analysis: The Case of S&L bankruptcy." *Financial Management*: 66-76.

Etheridge, H., R. Srinam and H. Hsu (2000). "A comparison of selected artificial neural networks that help auditors evaluate client financial viability." *Decision Sciences* **31**(2): 531-550.

Fernandez-Castro, A. and P. Smith (1994). "Towards a General Non-parametric Model of Corporate Performance." *Omega* **22**(3): 237-249.

Frydman, H., E. Altman and D. Kao (1985). "Introducing recursive partitioning for financial classification: The case of financial distress." *The Journal of Finance* **XL**(1): 269-291.

Gorzalczany, M. B. and Z. Piasta (1999). "Neuro-fuzzy approach versus rough-set inspired methodology for intelligent decision support." *Information Sciences* **120**: 45-68.

Greco, S., B. Matarazzo and R. Slowinski (1998). A New Rough Set Approach to Evaluation of Bankruptcy Risk. *Operational Tool in the Management of Financial Risks*. C. Zopounidis, Kluwer Academic Publishers: 121-136.

Guilhot, B. (2000). "Soixante-dix ans d'analyses théoriques et empiriques." *Revue Française de Gestion* **Septembre-Octobre**: 52-67.

Honjo, Y. (2000). "Business failure of new firms: an empirical analysis using multiplicative hazards model." *International Journal of Industrial Organization* **18**: 557-574.

Ignizio, J. and J. Soltys (1996). "Simultaneous Design and Training of Ontogenic Neural Network Classifier." *Computers Operational research* **23**(6): 535-546.

Izan, H. (1984). "Corporate distress in Australia." *Journal of Banking and Finance* **8**: 303-320.

Jeng, B., Y.-M. Jeng and T.-P. Liang (1997). "FILM: a fuzzy inductive learning method for automated knowledge acquisition." *Decision Support Systems* **21**: 61-73.

Jo, H. and I. Han (1996). "Integration of case-based forecasting, neural network and discriminant analysis for bankruptcy prediction." *Expert Systems With Applications* **11**(4): 415-422.

Jo, H. and I. Han (1997). "Bankruptcy prediction using case-based reasoning, neural networks, and discriminant analysis." *Expert Systems With Applications* **13**(2): 97-108.

Keasy, K. and R. Watson (1991). "Financial Distress Prediction Models: A Review of Their Usefulness." *British Journal of Management* **2**: 89-102.

Kim, C. N. and R. McLeod (1999). "Expert, linear models, and nonlinear models of expert decision making in bankruptcy prediction: A Lens model analysis." *Journal of Management Information Systems* **16**(1): 189-206.

Kiviluoto, K. (1998). "Predicting Bankruptcies With The Self-Organizing Map." *Neurocomputing* **21**: 191-201.

Koenig, G. (1985). "Entreprises en difficultés: des symptomes aux remèdes." *Revue Française de Gestion* **Janvier-Février**: 84-91.

Lacher, R., P. Coats, S. Sharma and F. Fant (1995). "A Neural Network for Classifying the Financial Health of a Firm." *European Journal of Operational Research* **85**: 53-65.

Laitinen, E. (1992). "Prediction of Failure of a Newly Founded Firm." *Journal of Business Venturing* **7**: 323-340.

Laitinen, E. and T. Latinen (1998). "Misclassification in bankruptcy prediction in Finland: human information processing approach." *Accounting, Auditing & Accountability Journal* **11**(2): 216-244.

Lee, K., I. Han and Y. Kwon (1996). "Hybrid neural network models for bankruptcy prediction." *Decision Support Systems* **18**: 63-72.

Lennox, C. (1999). "Identifying Failing Companies: A Re-evaluation of the Logit, Probit and DA Approaches." *Journal of Economics and Business* **51**: 347-364.

Leshno, M. and Y. Spector (1996). "Neural network prediction analysis: The bankruptcy case." *Neurocomputing* **10**: 125-147.

Lincoln, M. (1984). "An Empirical Study of the Usefulness of Accounting Ratios to Describe Levels of Insolvency Risk." *Journal of Banking and Finance* **8**: 321-340.

Luoma, M. and E. Laitinen (1991). "Survival Analysis As A Tool For Company Failure Prediction." *OMEGA* **19**(6): 673-678.

Mader, F. (1975). "Les ratios et l'analyse du risque." *Analyse Financière*: 8-18.

Makridakis, S. (1991). What Can We Learn From Failure, Insead, Working Paper, 26 p.

Mar Molinero, C. and M. Ezzamel (1991). "Multidimensional Scaling Applied to Corporate Failure." *OMEGA* **19**(4): 259-274.

Markham, I. and C. Ragsdale (1995). "Combining neural networks and statistical predictions to solve the classification problem in discriminant analysis." *Decision Sciences* **26**(2): 229-241.

Martin, D. (1977). "Early Warning Of Bank Failure: A Logit Regression Approach." *Journal of Banking and Finance* **1**: 249-276.

Martin-del-Brío, B. and C. Serrano-Cinca (1993). "Self-organizing neural networks for the analysis and representation of data: Some financial cases." *Neural Computing & Applications* **1**: 193-206.

McGurr, P. and S. DeVaney (1998). "A Retail Failure Prediction Model." *The International Review of Retail, Distribution and Consumer Research* **8**(3): 259-276.

Mensah, Y. (1984). "An Examination of Stationarity of Multivariate bankruptcy prediction Models: A Methodological Study." *Journal of Accounting Research* **22**(1): 380-395.

Meyer, P. and H. Pifer (1970). "Prediction of bank failures." *Journal of Finance* **XXV**: 853-868.

Olmeda, I. and E. Fernandez (1997). "Hybrid classifiers for financial multicriteria decision making: The case of bankruptcy prediction." *Computational Economics* **10**: 317-335.

Ooghe, H. and C. VanWymeersch (1986). *Traité d'analyse financière*, Presses Universitaires de Namur.

Piramuthu, S., H. Ragavan and M. Shaw (1998). "Using Feature Construction to Improve the Performance of Neural Networks." *Management Science* **44**(3): 416-430.

Rahimian, E., S. Singh, T. Thammachote and R. Virmani (1996). Bankruptcy prediction by neural network. *Neural networks in finance and investing*. Trippi and Turban, IRWIN.

Scott, J. (1981). "The probability of bankruptcy. A Comparison of Empirals predictions and Theoretical Models." *Journal of Banking of Finance* **5**: 317-344.

Serrano-Cinca (1996). "Self organizing neural networks for financial diagnosis." *Decision Support Systems* **17**: 227-238.

Shah, J. and M. Murtuza (2000). "A neural network based clustering procedure for bankruptcy prediction." *American Business Review* **18**(2): 80-86.

Sharma, S. and V. Mahajan (1980). "Early Warning Indicators of Business Failure." *Journal of Marketing* **44**: 80-89.

Slowinski, R. and C. Zopounidis (1995). "Application of the rough set approach to evaluation of bankruptcy risk." *International Journal of Intelligent Systems in Accounting, Finance and Management* **4**: 27-41.

Taffler, R. (1984). "Empirical Models for the Monitoring of UK Corporations." *Journal of Banking and Finance* **8**: 199-227.

Takahashi, K., Y. Kurokawa and K. Watase (1984). "Corporate Bankruptcy Prediction in Japan." *Journal of Banking and Finance* **8**: 229-247.

Tam, K. and M. Kiang (1990). "Predicting Bank Failures: A Neural Network Approach." *Applied Artificial Intelligence* **4**: 265-282.

Tan, C. (1996). A study on using artificial neural networks to develop an early-warning predictor for the credit union distress with comparison to the probit model. *Neural networks in finance and investing*. Trippi and Turban, IRWIN: 329-365.

Varetto, F. (1998). "Genetic Algorithms Applications in the Analysis of Insolvency Risk." *Journal of Banking and Finance* **22**: 1421-1439.

Vermeulen, E., J. Spronk and N. van der Wijst (1998). The Application of the Multi-factor in the Analysis of Corporate Failure. *Operational Tool in the management of Financial Risk*. C. Zopounidis, Kluwer: 59-73.

Vernimmen, P. (1978). "Panorama des recherches portant sur le risque du créancier." *Analyse Financière* **1er trimestre**: 54-61.

Zanakis, S. and G. Walter (1994). "Discriminant characteristics of US banks acquired with or without federal assistance." *European Journal of Operational Research* **77**: 440-465.

Zanakis, S. and C. Zopounidis (1997). "Prediction of Greek Takeovers Using Multivariate Analysis of Financials Ratios." *European Journal of Operational Journal* **48**: 678-687.

Zhang, G., M. Hu, E. Patuwo and D. Indro (1999). "Artificial neural networks in bankruptcy prediction: General framework and cross-validation analysis." *European Journal of Operational Research* **16**: 16-32.

Zimmer, I. (1980). "A Lens Study of the Prediction of Corporate failure by bank Loan Officers." *Journal of Accounting Research* **18**(2): 629-636.

Zmijewski, M. (1984). "Methodological Issues Related to the Estimation of Financial Distress Prediction Models." *Journal of Accounting Research* **22**: 59-82.

Zopounidis, C. (1985). Evaluation Multicritère du Risque de faillite d'Entreprises: Méthodologie et Application, Université Paris-Dauphine, Lamsade, Cahier N°85, 36 p.

Zopounidis, C. and A. Dimitras (1993). "The Forecasting of Business failure: Overview of Methods and new avenues." *Applied Stochastic Models and Data Analysis*: 1083-1107.

Zopounidis, C., A. Dimitras and L. Le Rudelier (1998). A Multicriteria Approach for the Analysis and Prediction of Business Failure in Greece. *Operational Tool in the Management of Financial Risks*. C. Zopounidis, Kluwer: 107-119.

Zopounidis, C. and M. Doumpos (1998). "Developing a multicriteria decision support system for financial classification problems: The Finclas system." *Optimization Methods and Software* **8**: 277-304.

Zopounidis, C. and M. Doumpos (1999a). "Business failure prediction using the Utadis multicriteria analysis method." *Journal of Operational Research Society* **50**: 1138-1148.

Zopounidis, C. and M. Doumpos (1999b). "A multicriteria decision aid methodology for sorting decision problems: The case of financial distress." *Computational Economics* **14**: 197-218.

On the Use of a Combination Approach to Automated Knowledge Acquisition Based on Neural Networks and Fuzzy Logic with Regard to Credit Scoring Problems

M. Michalopoulos[1], G. Dounias[2], D. Hatas[1], C. Zopounidis[1]

[1] Technical University of Crete
Department of Production Engineering and Management
Financial Engineering Laboratory
University Campus, Chania 73100, Greece

[2] University of the Aegean,
Department of Business Administration
Chios 82100, Greece

Summary:
This paper presents an application of automated knowledge acquisition in classification problems. The problem under examination is the classification of a sample of 130 enterprises into 5 different categories of credit risk. The data are both numerical and linguistic in nature, adding up to a total of 70 attributes. All data are obtained from an application form, which the company completes at a bank when seeking a loan and the only field which is provided by the bank is the final class. This classification task deals with statistical models, which try to give a clear view of the importance of every attribute by assigning a weight to it. Then, by multiplying each weight with the value of each corresponding attribute, the method produces a score. According to the range of the score the bank makes its decision. The problem, however, is that this method is very time consuming as it requires a long adaptation period in order to achieve the right results. The alternative solution presented in this paper is a combination of the fuzzy c-means clustering algorithm in conjunction with the Kohonen artificial neural network, which has a self-organizing Kohonen map. The basic idea behind this combination is to take advantage of the benefits offered by the two individual methods mentioned above and to compensate for each method's short-comings. The c-means process can be integrated into the Kohonen algorithm by replacing the learning rate with membership values thus combining the fuzzy c-means algorithm with the structure and adaptive rules of the Kohonen network.

Keywords: Fuzzy c-means, Neural networks, Credit scoring

1 Introduction

A tremendous rise in the use of different kinds of financial transactions has been observed in very recent years, both in banking and in the private business sector, a fact, which makes imperative the automation of some decision making processes. The most critical aspect of this automation is that it should take place without any high error rates during the data processing phase, while it should also require the least amount of time possible. Furthermore, the decisions taken with the use of an automated process should approximate as much as possible the corresponding decisions of an expert for the same problem. This is the reason why methods which imitate human thought and model human expertise and abilities have become very popular in the last two decades. Usually, all these "intelligent" methodologies for data analysis, decision-making, and model based reasoning, belong to the Artificial Intelligence approach [Simon (1991)]. Machine learning is that part of artificial intelligence's tools and techniques which seems to perform in the most promising way when dealing with real world applications [Langley & Simon (1995)]. Lately, various approaches have appeared in literature which function as knowledge generation approaches and could also be considered as hybrid computational intelligence approaches for real world application domains [Adrians and Zantinge (1996); Fayyad, Piatetsky-Shapiro, Smyth, Uthurusamy (1996)]. Particularly distinguished, as the reader should expect, are those applications dealing with financial domains and business applications [Nelke (2000)].

This paper deals with one of the most commonly used financial transactions in banking organizations, the financial application domain related to the granting of a loan. There are a large number of criteria, according to each case, which in practice, determine the restrictions and the final outcome of the decision process for granting a loan. For the fundamentals of credit scoring the reader should refer to [Lewis (1992)]. The subject of credit scoring is not so well-known in literature as one would expect [Thomas (2000)]. Most of the papers attempt statistical approaches for classifying candidates to be granted loans in two classes: positive and negative, see [Boyes, Hoffman and Law S.A.(1989); Thomas (2000); Thomas (1998)]. They usually perform discriminant analysis in order to produce a discriminant rule for these two classes [Steenackers and Goovaerts (1994)]. Other approaches use count data models in order to predict the number of times that an applicant for credit will not pay the accorded amount to return the credit [Guillen, Artis (1992)]. Other classic methods used to face the credit scoring problem include classification and regression trees, Markov chains, linear models and linear programming, graphical models, logistic regression, and many more [Thomas (2000); Thomas (1998)]. On the other hand, a limited number of intelligent methodologies has appeared over the last years focusing on credit scoring, such as neural networks [Desai, Convay, Crook and Overstreet (1997)], genetic algorithms [Desai, Convay, Crook and Overstreet (1997)], machine learning [Carter & Catlett (1987)], as well as data mining and Bayesian networks, see also [Thomas (2000); Thomas (1998)].

The present paper is not exactly the very first attempt at facing credit risk with the aid of computational intelligence techniques. Nonetheless, it is one of the ultimate approaches and is also quite sophisticated as it deals with a strong methodology which is the immediate result of the combination of the self-organized Kohonen neural networks and the fuzzy c-means clustering algorithm, equipped with new features, the best from each of the two methods and processes for learning from examples. On the other hand, special attention has been paid throughout this paper, to carefully model the application domain, a fact that contributes to a high degree to the final success of the overall attempt. Another clear aim of this paper is to introduce to banking experts a category of comprehensible new methodologies for handling a very difficult financial decision making problem. Especially in banks, which are the main loan-makers, there is a strong need for quick and valid evaluation of the applicant in order for some key decisions to be made. Some of these decisions concern whether to grant the loan or not, what its amount would be if granted, for how long its maturity should be and what the payback terms would be, in order to ensure the minimum possible risk for the bank. Several bank executives devote much time to this kind of decision-making in an effort to give a scientific and fully understandable way of dealing with the problem which at the same time, must fulfill all the requirements set by the bank for the risk level and provide for speed and clarity of the decisions.

The paper is organized as follows: The basics of the credit scoring application (CSA) are presented in section 2. Then, in section 3, a clear view of the data constituting the problem is given as well as the way in which they are modeled and organized. In section 4, follow the basic theory of both the Kohonen neural network in addition to the fuzzy c-means clustering algorithm and in section 5, the proposed overall methodology is presented and the acquired results are discussed. Finally, in section 6, concluding remarks drawn from the presented approach are given.

2 Credit Scoring

Before the presentation of the followed methodology and of the actual application, it is imperative to give a brief presentation of the general concept behind the term Credit Scoring. The basic concept behind the methodology of CSA is that the near future has a high degree of similarity to the near past. One can come to this conclusion even by using one's everyday experience; with every new day, the things expected to occur as much in regular business as in personal life, do not differ significantly from those which occurred the previous day or days. If this conclusion is applied in the field of credit risk assessment, where a bank or any other loan maker always tries to grant from a loan given to a new customer, it is quite certain, that each new potential customer will behave similarly to a previous one having the same characteristics. For instance, if the customers' demographic and financial characteristics demonstrate a substantial degree of similarity, then all the necessary requirements to reach the above conclusion are met. This projection of the near past onto the near future is usually accomplished by using proper statistical models [Thomas (2000); Thomas (1998)], such as multiple regression and discriminant analysis, the results of which are valid only if a set of demands and criteria are met.

CSA is a process during which the potential client is graded, with the use of the proper statistic methods [Lewis (1992)]. This is accomplished by multiplying the various kinds of information with a different weight according to the field and summing up at the end. This score represents a specific possibility for the potential client to behave in a certain way according to a clearly stated criterion given by the bank. The information about the client is gathered mainly from the loan application filled in at the time of the request, from previous data that the bank maintains about that client or similar types of clients and finally from an outside information database. The final outcome of the entire process is to automatically reject or approve the application, without the need of the human factor, after having taken into consideration the total of the individual score that the client has accumulated and the cutoff base predetermined by the bank according to its demands.

The history of credit scoring is relatively short and began in the US. Many executives had tried to find a scientific approach to an evaluation process for credit risk applications based on quantitative criteria, but all efforts had been unsuccessful since there were not any computers that had the processing power of today's personal computers. Henry Wells, an executive of Spiegel Inc., may well be considered a pioneer in this field. During the Second World War, he used a number of statistical techniques to form a credit scoring system for the purpose of replacing the credit analysts whom companies used at that time. Many experts tried now and then to develop similar methods, but the appearance and activation of the Fair Isaak firm was undoubtedly the key point.

The basic setback in the evolution of credit scoring lay in a lack of conviction in it on the part of executives specializing in the field of credit who regarded the system as one whose sole purpose was to leave them without a job. However, the circumstances that followed in the decades of the 70's and 80's forced all the major financial organizations to use quantitative techniques in order to assess the credit risk at hand. One problem that has still remained unsolved throughout the process of creating the credit scoring system, is that concerning the use of the data representing the cases already rejected, whose behavior had they not been rejected is not known.

The use of the aforementioned quantitative methods has recently spread in banking. Its main application areas range from a basic evaluation of credit risk for a potential client using the data gathered from his application to a wide variety of products that covers all the forms of consumer credit and generally every financial activity. On the Internet one can find information[1] ranging from specialized software for a specific application of credit scoring, to instructions on how to achieve great score numbers in order for his application for a consumer credit product to be eligible for grant.

3 Presentation of the Problem

In this section a complete description of the problem of granting a loan is given, while the application area and the variables which constitute the data set are also presented.

3.1 Description of the problem

The problem lies in the bank's decision-making process, as to whether or not a loan should be granted to an enterprise and with what amount of certainty, based only on the data gathered from the application of the potential client. The bank from which the data were gathered, required the following:

- A bank branch application form completed by the potential client. The form included some general fields providing information not available through normal financial statements such as a balance sheet. One example of this type of information regards the particular kind of software the company uses. Another example deals with the number of people in the company's employ. Yet another concerns the main activities of the company.
- A balance sheet.
- An income statement.
- Additional information gathered from information agencies by the bank executive. This information includes what the firm's position is among its peers, its potential, the level of skill of its executives and other similar information.

To sum up, the experimental data were gathered from four different sources and included both numerical data in the form of percentages, indexes, or simple numbers, and qualitative data consisting of terms. In the classic system used by banks to date, the qualitative data are transformed into numerical with the use of certain transformation scales. Here lies one of the major disadvantages of conventional methodologies which try to convert linguistic terms and concepts into numerical equivalents thereby losing the essence of the words.

---

[1] See the site www.creditscoring.com

3.2 Application area

The application area, is none other than the relevant department of any financial organization which grants loans to firms. These same financial organizations employ highly qualified experts, who possess both a great deal of field experience and a firm knowledge of the problem at hand, i.e., the granting of a loan to firms which issue accounting books of specific categories.[2] At long last, there is a need for a computer system with increased processing power allowing for the safe and quick processing of data.

3.3 The data

In order to achieve the goal of managing credit scoring with the use of computational intelligence techniques, a set of data which fulfill the presuppositions below, should be available:

- Represent the general population
- Are homogenous
- Have as little as possible noise
- Should extend over all fields (although inductive learning deals with any empty fields)
- Should include be a substantial number of cases
- Should cover all the different classification categories

Keeping in mind all the above, the phase of gathering data followed and concerned data that were involved in the bank's decision whether or not to grant a loan to a firm which keeps accounting books of specific categories. The bank examined in this paper has been using a classic application credit scoring system. The system keeps as input, 70 fields in total. For each of the 130 cases in which a loan had been granted in the past by the bank in question, there were 70 attributes covering a period of up to three years before each loan was granted. Below, an analysis follows of the basic elements of the specimen allowing the bank to classify the cases into one of the following classes: "Risk-free", "Average Risk", "Under Surveillance", "High Risk", "Weak" (i.e. impossible to receive a loan).

4 Proposed Methodology

The proposed methodology consists of simple steps. The first step is to build the knowledge database. In order to accomplish this the necessary data need to be gathered and can be found in two different forms: data printed on paper and data

---

[2] In Greek accounting system these categories correspond to A and B class books.

accessed from electronic databases. The data may not always be in a form which the computational intelligence can handle, so a transformation would have to take place to ensure the compatitability of the constructed database. The database in its turn will provide the training base of a Fuzzy Kohonen neural network, for the sole purpose of extracting knowledge. The extracted knowledge will then be automatically transformed into an acceptance rule database for new cases. Before taking the next step, it is imperative to ensure the effectiveness of the rule database at this stage. In most cases, several training scenarios must be implemented to ensure that all cases are represented in the input file and that the network has extracted efficient rules from them. When this has been established, one can take the next step.

The next step is to link the rule database to the online database through a software implementation, and, as soon as the data of a new case are filled in, to provide an assessment of the case, either accepting or refusing the applicant's request for a loan. Again a transformation process, which is the same as in the training phase, will have to take place. The results will then be presented to the user. At this point, it is preferable for the software to accept human intervention in the assessment, in order for an individual to rectify some potentially faulty assessments.

The above-described process of the proposed framework can be visualized in the process flowchart of Figure 1.

5 Computational Intelligence Methodology

In this section the basic theorics of computational intelligence, which in combination provide the proposed architecture, will be presented. To begin with, there will be a short introduction of Fuzzy Logic, which was the first non-conventional technique used in the industry to solve problems in the general sense of the word. Next, the theory of Neural Networks will be briefly mentioned. Finally, at the end of this section, the theory of Fuzzy Kohonen networks will be presented in which, in our case, the problem of Credit Scoring found a suitable solution.

5.1 Fuzzy Logic

Over the years the classical methodologies used to adequately solve most of the problems became redundant. As a result, other non conventional methodologies have been created in order to deal with problems with greater success. Fuzzy logic is the first non-conventional technique used in the industry for solving problems and was presented by Zadeh in the middle of the past decade [Zadeh (1985); Zimmerman (1996)]. As is well-known, fuzzy logic is based on the assumption that the surrounding environment consists of data belonging to different data sets with different degrees of membership. The most common definition is:

For X a crisp (classical) set of objects, which is to be analyzed by fuzzy means, then $\tilde{A} := \{(x, \mu_{\tilde{A}}(x));\ x \in X\}$ represents a fuzzy set of X, where $\mu_{\tilde{A}}$ represents the membership function and $\mu_{\tilde{A}}(x)$ represents the degree of membership of element x to the fuzzy set $\tilde{A}$. As a fuzzy number we take every convex, fuzzy set, normalized over the interval [0,1] of the set of real numbers. Precisely one element with a degree of membership 1 exists. The membership function of this fuzzy set is piecewise continuous. Finally, as a linguistic variable is considered every variable whose values are not numbers, but rather linguistic constructions. The contents of these terms are defined by fuzzy sets over a base variable.

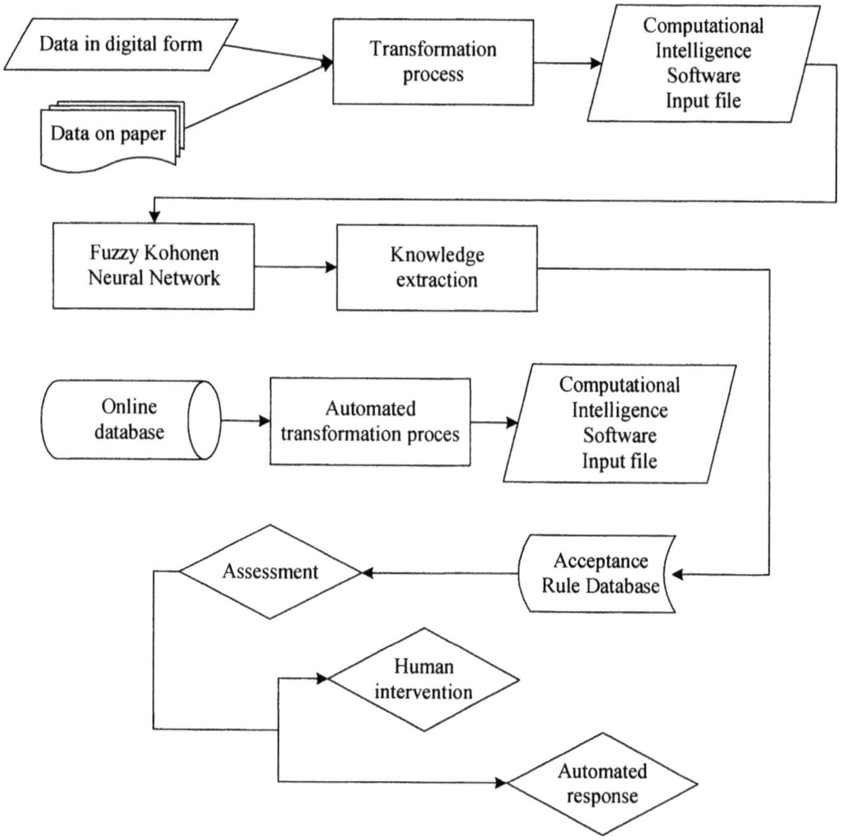

Figure 1: Process flowchart of the proposed methodology

The complexity of the theory is the major reason for its not being applied sooner in the field of industry. However, the applications showed that it is a strong tool in the hands of an expert, not only for the purpose of representing knowledge and experience, but also for creating conclusion extraction mechanisms which use the available encrypted knowledge and the current value of the attributes of the process under control to extract the required control actions.

5.2 Neural Networks

Another theory that is gaining more and more support is that of Neural Networks [Kosko (1992); Lin (1994); Freeman (1991)]. The new idea behind the theory of artificial neural networks is to try to imitate the structure and workings of the human brain by means of mathematical models. The foundations of most neural networks are based on three basic qualities of the human brain: knowledge is distributed over many neurons within the brain; neurons can communicate with one another; and the brain is adaptable. The terms used to describe neural networks are derived from these three qualities of the human brain and are: I) Structure of the Neuron, II) Network Topology and III) Adaptation (learning) Rule.

The neurons, or processing units, which make up the neural network, are single elements that consist principally of the four following components: a connection function, an input function, an activation or transfer function and an output function. A neuron receives a signal via several input connections. These are weighted at the input to a neuron by the connection function. The weights employed here define the coupling strength (synapses) of the respective connections and are established via a learning process, in the course of which they are modified according to given patterns and a learning rule. The input function compresses these weighted inputs into a scalar value, the so-called network activity at this neuron. Simple summation is generally employed here. In such cases, the network activity, which results from the connection function and the input function, is the weighted sum of the input values. The activation function determines a new activation status on the basis of the current network activity, if appropriately taking the previous status of the neuron into account. This new activation status is transmitted to the connection structure of the network via the output function of the neuron, which is generally a linear function. There are various types of activation functions; the type used often depends on the learning rule.

As far as the topology is concerned, a layered connecting structure is generally employed, whereby the layer on which the input signals act is referred to as the input layer the layer at which the results are collected is known as the output layer and the layers located between these are known as hidden layers. The neurons are generally fully connected on a layer-by-layer basis. The number of layers often determines the performance of a network. A distinction can be made between feed forward, lateral and feedback connections for the method of linking the different layers. Both feed forward and feedback connections over several layers are conceivable.

The connecting structure and the choice of processing units determine the structure of a network. In order to carry out classification, the network must be taught a task by presenting it with examples in a training phase. The knowledge relating to this task is not stored in a knowledge base, but is distributed throughout the network in the connection weights. The weights are adjusted in this phase to fit the problem. This is done by means of learning rules or algorithms. The training phase normally proceeds as follows: random values are initially assigned for the weights of the neurons. Patterns from a training data record are then presented to the network and the weightings are adapted on the basis of the learning rule and the training pattern until a convergence criterion is attained.

The learning process can be divided into three different types. The first is that of supervised learning, where in addition to the input patterns, the desired corresponding output patterns are also presented to the network in the training phase. In the case of unsupervised learning, the network is required to find classification criteria for the input patterns independently. Lastly, in the case of stochastic learning methods, random processes and probability distributions are employed to minimize a suitably defined energy function of the network.

Our approach utilizes a software tool[3] which incorporates the above theories. More specifically, it offers the following methods for the actual obtaining of information: Fuzzy Rule Based Systems, Multilayer Perceptron, Fuzzy C-Means, Kohonen Network and Fuzzy Kohonen Network. For each of the above methods, there is an editor with which models can be prepared and processed. The initial data can be read into the software tool from a number of different sources. In addition to the software's own file format, data can be imported from ASCII files, Microsoft Excel files, as well as from databases. The ODBC interface provides an efficient means of accessing data from different types of databases. It also provides a graphic macro language in order to help the user automate sequences of processing steps and to give the user the ability to intergrade new functions.

5.3 Fuzzy Kohonen Networks

A fuzzy Kohonen clustering network (FKCN) [Bezdek (1992)] is a neuro-fuzzy model in which fuzzy sets and artificial neural networks, rather than being stand-alone models, exist side by side in one complete model. Specifically, a self-organizing kohonen map [Kohonen (1989)] is combined with the fuzzy c-means algorithm [Bezdek (1981)], both of which have been discussed in the previous

---

[3] The software tool used was DataEngine. A product of MIT, Germany. Additional information can be found at the product's home site in the internet at http:\www.mitgmbh.com

chapters. In more recent publications, the process has also been called fuzzy learning vector unitization.

The goal of combining fuzzy sets and neural networks is to take advantage of the benefits offered by each individual method and to compensate for each method's short-comings. Self-organizing maps in particular have algorithmic weakness, which can affect the learning process for the worse:

- parameters for the learning process such as the learning rate or the neighborhood size for the adjustment of the weights, are determined only by heuristic means. A poor choice of parameter not guaranteeing the success of learning;
- the result of the training process depends upon the presentation order of the training examples, since the weights are adjusted after the presentation of each item.
- the convergence of the learning process is not based on optimization but rather is forced by the decreasing neighborhood size and learning rate. Hence, it is not guaranteed that the network can learn the given problem and represent its complexity fully.

Since the c-means process, as opposed to Kohonen network training, is an optimization algorithm, it can be integrated into the kohonen algorithm as a way of eliminating the Kohonen network's short-comings. The idea behind this is that the learning rate is replaced by membership values [Huntsberger 1992] and in this way the fuzzy c-means algorithm is combined with the structure and adaptive rules of the Kohonen network.

6 The Application

The fuzzy Kohonen network is a Kohonen network whose training method has been improved on considerably by integrating the fuzzy c-means algorithm. It is thus one of the so-called neuro-fuzzy methods. The main applications of the fuzzy kohonen network are classification tasks. The development of a classifier on the basis of a fuzzy kohonen network can be divided into four phases:

- Training
- Labeling
- Test / Validation
- Use of the model (Recall)

In the training process, the weight vectors of the neurons of the fuzzy Kohonen network were adapted by means of the training examples in such a way that the neural network supplies as good a representation as possible of the training data set. The weight vector of a neuron of a fuzzy Kohonen network corresponds to the feature vector of an item in the feature space under study and the training process calculates a set of characteristic items for the training data set. In a second supervised step, labeling, cluster labels are assigned to the individual neurons.

This is done via the interpretation of the content of the weight vectors (feature vectors) of the neurons. Here the same name can be assigned to several neurons. Thus, a cluster is represented by several neurons in the fuzzy Kohonen network. After validation of the fuzzy Kohonen network by a series of test items, it can be used as a very powerful classifier.

The general procedure for the creation of a fuzzy Kohonen network classifier as described in the manual of the software tool is as follows:

- Configuration of the classifier
- Setting of the training parameters
- Specification of the input/output files
- Training of the classifier (training of the network using the input items)
- Labeling (labeling of the neurons using example items)
- Validation of the classifier via a test
- Recall (classification of the input items)

For the configuration of the classifier, it is necessary to set suitable parameters. These include the number of features to be examined, the number of dimensions and the neurons per dimension. It is also necessary to specify parameters for training. These determine the initialization of the neuron weights as well as different training options such as exponent and stop condition.

Various files are used in the different phases of training, validation and execution of the classifier. A fuzzy Kohonen network receives its input information from the input files, and the results of the steps involved are saved in the output files. After proper configuration, the network is trained. Using example items, the neurons are given suitable labels in the so-called labeling process. The suitability of the classifier can be checked by means of a test. If it is satisfactory, the classifier is used in the classification phase. Otherwise, it could be necessary to adapt the classification or training parameters and repeat the training steps.

The input for the software is a datasheet much like that of Microsoft Excel. Each line represents a new case, while each column is an attribute. The drawback was that we could not enter linguistic values in the cells, so we had to produce a suitable transformation of the linguistic values into numerical ones. We believed that the most common thing to do when facing yes or no answers was to assign the values 1 and 0 according to the significance of the answer. For instance, if the desired answer for the bank was a 'no' then 'no' will have to take the value 1. For the circumstances where there were more than two linguistic variables we applied a scale of 0 to 10, trying to give a suitable meaning to the words at hand. As regards the numerical data, we used the built-in normalization tool of the software. Finally, after all the necessary transformations were completed the input file was ready.

For training, another file was required. This file was an augmented input file where there were as many additional columns as there were classes. In a particular line, one could find the value 1 only in the column of the corresponding class and the value 0 in the others. This file was needed in order to provide for the labeling input file which is part of the supervised learning process.

The number of attributes was set at 70, the data input file in our case. Next, a selection for the dimension of the inner map of the neural network had to be made. According to the theory of neural networks, a two dimensional map would be adequate if in the first layer, we have a number of neurons equal to the number of attributes and if in the second dimension we have a number of neurons equal to the number of classes. The numbers for our case were reduced to a two dimensional map with 70 neurons in the first dimension and 5 in the second.

For the neuron connection weight assignment process, we chose to have a random seed with a lower boundary of 0 and an upper boundary of 1. The exponent was defined by means of trial and error. After several test trainings, we discovered that as an initial exponent, it was preferable to use the value 1,08 which provided us with the best results. The value of the step was set at 0,0001. In conclusion, the parameter defining the end of the training, also known as the convergence threshold, was also set at 0,0001.

The entire training process was not at all time consuming and completed in 12 steps. At this point we would like to stress out that initially we used only 80% of the available database, creating therefore a holdout sample of 20%. The results where more than promising and we decided to include the maximum available cases in the training procedure, giving us the means for comparison. As it turned out the error rate did not vary significantly. Due to this fact a decision was made to use all the sample cases so as to create the most complete initial rule database. After all we are in the process of gathering new cases so as to re-validate the model. To end, for the 130 test cases inserted in the data input file, we had an error rate of 0,8%, which means that the extracted knowledge database gave the wrong class for only 1 case. The training cases were gathered from a variety of cases and were representative of almost every subcategory of the bank's clients and the almost negligible error rate is truly our justification for the proposed framework. The astonishingly low error rate also implies that the Fuzzy Kohonen network has very good adaptation skills, a feature which we believe will prove to be a valuable asset when making a real application, which has to deal with large databases.

7 Concluding Remarks

In this paper we have presented an application of automated knowledge acquisition in the field of credit scoring. The results of this implementation utilizing Fuzzy Kohonen neural networks are more than promising. Through the astonishing error rate of 0,8 % we have at least proven the effectiveness of the proposed framework. The whole process of decision-making can be automated and decisions could then be made in a matter of a few seconds.

The potential for the use of computational intelligence techniques in the field of credit scoring is enormous, not only due to the application of such methodologies in financial problems, but also due to the excellent results they tend to provide. What we have presented is a realistic approach to a solution which meets with the preferences of the decision maker. The major advantages which we believe will urge the financial sector to adopt such techniques are that of a minimum time requirement to establish the initial rule database, the ease of data manipulation and finally, the ability to revise the extracted knowledge database at any given point.

At the present time, we are beginning a data gathering process in order to create a new testing database, so as to reevaluate the current results. Furthermore, we are planning to apply other computational intelligence techniques such as evolutionary computation and some genetic programming approaches.

References

Adriaans, P., and Zantinge, D., (1996), Data Mining, Addison-Wesley Public

Bishop, Chris, (1995), Neural networks for pattern recognition

Bose ,N. K, (1996), Neural network fundamentals with graphs, algorithms, and applications

Boyes, W.J., Hoffman, D.L., and Law S.A., (1989), "An Econometric Analysis of the Bank Credit Scoring Problem", Jour. of Econometrics, Vol. 40, pp.3-14.

Carter, C., & Catlett, J., (1987), "Assessing credit card applications using machine learning", IEEE Expert, Vol. 2, pp. 71-79

Desai, V.S., Convay, D.G., Crook, J.N., and Overstreet, J.A., (1997), "Credit scoring models in the credit union environment using neural networks and genetic algorithms", IMA Journal of Mathematics Applied to Business & Industry, Vol. 8, pp. 323-346

Fayyad, V.M., Piatetsky-Shapiro, G., Smyth, P., Uthurusamy, R., (1996), Advances in Knowledge Discovery and Data Mining, AAAI & MIT Press

Freeman, James A., (1991), Neural networks

Guillen, Artis, (1992), "Count Data Models for a credit scoring system", Third Meeting on the European Series in Quantitative Economics and Econometrics on "Econometrics of Duration, Count and Transition Models", Paris, Dec.10-11, 1992, pp.1-9

Herbert, Simon, (1991), "Artificial Intelligence: Where Has It Been and Where is it Going?", IEEE Trans. Knowledge & Data Eng., Vol. 3, No. 2, pp. 128-136

Huntsberger, T.L., (1992), Parallel Self-Organizing Feature Maps for Unsupervised Pattern Recognition, IEEE Press, New York, pp.483-495

Kohonen, Teuvo, (1989), Self-organization and associative memory, Springer - Verlag, New York

Kohonen, Teuvo, (1997), Self-organizing maps, New York, Springer

Kosko, Bart, (1992), Neural networks and fuzzy systems

Langley P. & H. Simon, (1995), "Applications of Machine Learning and Rule Induction", Communications of the ACM, 38 (11), pp.55-64

Lewis, E.M., (1992), An Intoduction to Credit Scoring" Athena Press, San Rafael, CA, USA

Lin, C. T., (1994), Neural fuzzy control systems with structure and parameter learning

Mitchell, T.M., (1997), Machine Learning, McGraw-Hill, New York

Nelke Martin, (2000), "Business Intelligence in Financial Engineering – A combined Approach to Customer Segmentation and Database Marketing", in Proc. of CoIL-2000, European Symposium on Computational Intelligence and Learning, 22-23 June 2000, Chios, Greece, pp. 171-177

Nomikos, I., Dounias, G., Vemmos K., (1999), "Comparison of Alternative Criteria for the Evaluation of Machine Learning in the Medical Diagnosis of Stroke", in Proc. 3rd Int. Data Analysis Symp., Aachen, Germany, pp.63-66

Quinlan, J.R., (1986), "Induction of Decision Trees", Machine Learning, Vol. 1, pp. 81-106

Quinlan, J.R., (1993), C4.5: Programs for Machine Learning, San Mateo: Morgan Kaufmann

Soucek, Branko, (1991), Neural and intelligent systems integration, IRIS Group, 1991

Steenackers, A. and Goovaerts M.J., (1994), "A Credit Scoring Model for Personal Loans", Insurance: Mathematics and Economics, Vol. 8, pp. 31-34

Thomas, L.C., (1998), "Methodologies for classifying applicants for credit", in Hand, D.J., & Jacka, S.D. (Eds.), Statistics in Finance, Arnold, UK, pp.83-103

Thomas, L.C., (2000), "A survey of credit and behavioural scoring: forecasting financial risk of lending to consumers", International Journal of Forecasting, Vol. 16, pp. 149-172

Tsoukalas, Lefteris H., (1996), Fuzzy and neural approaches in engineering

Von Altrock, Constantin, (1997), Fuzzy logic and neuro-fuzzy applications in business and finance

Von Altrock, Constantin, (1995), Fuzzy logic and neuro-fuzzy applications explained

Vonk, E., (1997), Automatic generation of neural network architecture using evolutionary computation

Weiss, S.M., N. Indurkhya, (1998), Predictive Data Mining: A Practical Guide, Morgan Kaufmann

Zadeh, L. A., (1988), Fuzzy logic, Computer, Vol.21, No.4

Zadeh, L. A., (1965), Fuzzy Sets, Information & Control, Vol.8

Zimmerman, H. J., (1996), Fuzzy Set Theory and its Applications, Kluwer Academic Publishers Boston / Dordrecht / London

Banking Organization, Banking Marketing and Customer Relationship Management

Bank-firm Relationships: A Review of the Theoretical Literature

M. Troege

ESCP-EAP
Département de Finance
97 av. de la République
75543 Paris, Cedex, France

Summary:
This paper reviews the microeconomic literature on relationship banking. We classify the existing work, explaining the main ideas underlying the different approaches and point out the implication for bank management as well as for bank regulation. Particular attention is given to credit market competition. Empirical literature is only reviewed as far as it is directly related to a theoretical concept.

Keywords: Relationship banking, Screening, Monitoring, Credit market competition

1 Introduction

Bank loans have the same payoff profile as commercial paper or corporate bonds. In both cases the borrower has to pay a fixed interest charge and can be forced into bankruptcy whenever it fails to meet these obligations. However bank loans and bonds are definitely not perfect substitutes: Firms that have access to bank credit do not necessarily have access to the bond markets and quite often firms are financed with junk bonds, which would have severe difficulties in finding bank credit.

What makes banks loans and bonds different? It is surprisingly difficult to give a convincing and complete answer. Apparently it is not only the payoff profile, which determines the nature of a financial contract but also the identity of the provider of finance. Whereas credits are held by one or at most a few banks, bonds are usually owned by a large number of dispersed and mostly unorganized private investors. In order to understand why this changes the nature of the financial contract one has to look more closely into the mechanisms of bond and credit markets.

In fact, bond markets are not so difficult to understand. They are a fairly good example of almost perfect Arrow (1964)-Debreu (1959) markets. Market power and asymmetric information between investors do not usually play an important role. In addition the investment of each bondholder is usually small compared to the size of a company so that bondholders have very low incentives to influence the decisions of the company. Once they have bought the credit they will simply wait and see whether the firm will repay. They will rather try to reduce risk by diversification than by intervening into he firm's policy.

More than 40 years of research have provided an extensive understanding of the mechanisms of these simple markets. However, all the assumptions of perfect markets are obviously violated when it comes to bank credits: Banks have usually market power when dealing with a firm, they also have inside information about the firm's project and its quality and finally they have the incentives and the bargaining power to implement changes in the firm's policy.

This makes credit markets much more difficult to understand than bond markets. Information and incentive problems are abundant. Only in the last decade the advance of information economics and game theory has made it possible to analyze the relationship between banks and their customers in a formalized way and address some of the particularities of credit contracts. Not surprisingly the results are much less clear-cut and intuitive than in the case of bonds. However, a number of insights can be gained from this literature, which are not only of purely theoretical interest but which have important consequences for the management of banks as well as for the design of banking and credit market regulation.

This article will give a short and synthetic review of the main strands of the literature on banking relationships, underlining the implication for bank management and regulatory policy. We will point out the major tasks of a bank in the next section describe how these tasks are affected by competition in Section and point out the implications for bank management and policy in Section 5 and 6.

2 The Added Value of Banks

Efficient financial markets in general and banks in particular have to carry out several functions. Similar to other markets, efficiency requires that capital be provided at a price close to marginal costs. However, unlike in usual product markets, the allocation of capital is not only determined by its price. Banks choose their customers and play therefore an active role in allocating savings among alternative investment uses. Welfare is only maximized if they direct funds to the firms with the highest returns. In addition, the provision of finance is not a spot transaction but extends over a period of time during which the provider of finance can influence the borrower. Again efficiency requires that the provider of finance find ways to ensure that funds are not squandered.

Hence banks have to fulfill at least three main tasks: They should provide cheap finance, they should allocate the funds to the firms with the highest returns and they should provide efficient corporate governance. Simple efficiency of the internal management processes is only necessary to keep the spread down - nevertheless most of the banks' efforts concentrate on this aspect. However, there is clearly a tradeoff between low costs and efficient allocation and control of finance. Clearly identifying good borrowers and supervising credits is costly. Low spreads make no sense if bad firms are financed or if firms are badly supervised.

In the following paragraphs we will discuss in more detail the allocation of credit as well as the credit supervision and point out the importance of information and information acquisition for executing both tasks properly.

2.1 Allocation of finance

In order to identify good borrowers the bank needs information about the quality of the borrower. They conduct creditworthiness test, talk to the managers and they analyze the general prospects of the firm. Economists use the term "ex ante monitoring" or "screening" to characterize this activity. It is clear that this activity is costly and therefore the amount of information collected will depend on the incentives of banks to acquire this information. The information acquired before giving a loans is crucial for the allocation of capital in an economy. If banks have insufficient incentives to screen, they will refuse to finance complicated or innovative businesses the creditworthiness of which is difficult to judge.

The factors influencing the incentives of banks to screen borrowers have been analyzed by a number of papers. Chan, Greenbaum and Thakor (1986) show that the screening effort of banks decreases with the reusability of information, hence will be low in fast changing, risky environments. Manove, Padilla and Pagano (1998) show that the presence of collateral may lead to an inefficiently low screening level. In addition, banks will not acquire the socially relevant information. For example, it is to be suspected that they will be more interested to know the default probability than the maximal return in a best case scenario.

The most important factor influencing screening activity is, however, competition. The impact of increased competition on the incentives to screen borrowers is far from evident and probably ambiguous. On the one hand more intense competition could erode incentives because of free- riding effects. A bank could observe the information collected by its competitor and use it to make a competing credit offer. On the other hand it is clear that screening can also give competitive advantages if the information can be kept secret. Intense testing could enable banks to discover potential clients that have been neglected by other banks. Section 3 will analyze in more detail the approaches that have been used to shed light on this question.

2.2 Corporate governance

Banks do not only acquire information before they provide a loan; they also check the firm's business while the loan is outstanding. This activity of credit supervision has been termed "monitoring" by economists. Whereas screening is not unique to banks – at least to some extend bondholders will also collect information about the firm before buying a bond - monitoring is an activity specific to banks.

The kind of information that will be useful for the bank depends its possibilities to influence a firm. It is evident that banks would have no incentives to monitor loans if they could not prevent the failure of repayment in case they have discovered a problem at the firm. In fact, in this case, the are in the situation of a bondholder. In most cases, however, banks will be able to react to a problem at the firm by withdrawing its credit or increasing collateral. In this case it will be sufficient for the bank to know the firm's business well enough to anticipate distress situations early. Only if the bank has more extensive channels of influence, for example through a post in the supervisory council, it will be useful to develop precise knowledge of how to conduct the firm's business, influence personnel decisions etc..

In the United States credit contracts are very detailed and do not give a bank many possibilities to influence the firm one the loan is given. In addition, as argued by Kroszner and Strahan (1999) legal liability makes it very difficult for banks to influence firms for example through outside directorships on the firm's board. Basically banks can only intervene when the firms violates a pre-specified covenant and in this case it can not do much more than simply call back the credit. Rajan (1995) has modeled how covenants and collateral may give a bank incentives to monitor and analyzed under which conditions banks implement efficient liquidation.

In contrast to the United States credit contract in Europe tend to be much more flexible and at the discretion of the bank. For example almost all long-term credit contracts in Germany contain a so-called "material adverse clause". This clause enables the bank to call back the credit or even increase interest rate without specifying clear conditions. It is clear that this possibility gives the bank much higher incentives to control and influence the firm. At the same time banks frequently occupy positions in the supervisory board which provides them with privileged information about the firm's health and its strategy.

A number of papers have argued that this influence of the bank is beneficial to the firm. For example Carminal and Matutes (1997) assume that after having invested in monitoring, banks can directly influence the decisions of the firms and will do this in a socially beneficial way. In Holmström and Tirole (1997) monitoring by a bank alleviates moral hazard by decreasing the private benefits of risky behavior.

However it is clear that too much bank influence may also distort the firm's decision making. In particular banks will try to implement low risk strategies instead of encouraging strategies, which create high expected welfare but are maybe riskier. For example Garella (1996) argues that the rents generated by banks can be explained by the bank's reduction of a borrower's riskiness.

Most importantly the ability of one bank to understand and influence firm will again have an impact on competition. Banks that have not monitored the firm will incur a higher credit risk. Monitoring on the other hand is not profitable for an outsider, as he will never recoup his monitoring costs. Hence ex- post competition is reduced. This idea of ex post rent extraction has been pioneered by Sharpe (1990) and further exploited by Fischer (1990) and Rajan (1992). Again we will refer the reader to Section 3 for a detailed discussion of these issues.

3 Relationships and Competition

It is evident that a certain extent of competition between banks is necessary to achieve efficient financial markets. However, strong competition between banks has not always been regarded as a good thing. The main concern has been that excessive competition endangers the stability of the financial system. As the financial system has the character of an infrastructure, the failure of banks can have severe consequences for an economy. However, economic stability is not the only reason why competition could be harmful. The above discussion has made clear that the effect of information on competition and the effect of competition on the incentives to acquire information is a complex issue. A large literature has evolved which tries to untangle the different mechanisms involved. We will review this literature in the following section, abstracting from the question of how competition affects stability.

3.1 The industrial organization approach

Bank representatives like to claim that money is a very homogenous good and that therefore the credit market is necessarily perfectly competitive. Indeed most of the early banking literature uses the Bertrand model with identical marginal costs (refinancing rates) where banks do not make any profits. However, banks clearly make profits. One of the earliest models of bank behavior, the Monti-Klein model (Klein 1971, Monti 1972) avoids this paradox by assuming monopolistic banks. A more realistic model of banking competition, which also generates nonzero rents, is the generalization of the Monti-Klein model to a two-sided Cournot oligopoly (compare Rochet/Freixas (1997 p. 59). In fact, Yanelle (1989) and (1997) has shown that nonzero rents and inefficiency can also be the result of two-sided Bertrand competition, where banks first compete for deposits and then on the credit side.

Another possibility to avoid the Bertrand paradox, which has become popular in the banking literature, is the Salop (1979) spatial competition model. It has been used for modeling competition on the deposit side (Matutes and Padilla (1994), Cordella and Levy-Yeyati (1998)) but also on the credit side (Chiappori et al. (1995), Gehrig (1996)). Indeed, the Salop model is probably a good description for what happens on the retail and depository side of the business. Retail customers have to pass frequently at a bank for example to get money and tend therefore to choose a bank close to their home. However, product differentiation is probably not a good explanation of the economic mechanism that allows a bank to earn rents in the corporate loan markets. Normally, banks are sufficiently close to each other for the travelling costs to be negligible compared to the amount of credit provided to a company.

It has been argued that differentiation should be understood in a more abstract sense for example as differences in services provided by the bank. It is however not clear whether this kind of differentiation can be captured with a simple spatial competition model. In fact services like consulting are rather an additional product sold by the bank. Why these additional products generate rents in the credit business would have to be explained with a detailed model..

3.2 Screening and the auction approach

We have argued above that the main reason for imperfect competition, in the credit markets is not differentiation in the usual product market sense but that imperfect competition arises because of informational asymmetries between banks. In fact, considering banks, in analogy to usual firms as producers of credit and borrowers as consumers misses a crucial difference between credit and product markets: Producers do not care about who buys the product, whereas banks care very much to whom they give the credits.

In fact understanding the loan as the price a bank pays to obtain the right to receive an uncertain future cash flow may be a more appropriate way of looking at banking. This is an approach closer to the theory of financial markets, where the firm sells bonds to the investors. Unlike in financial markets, however, as usually there are only a few banks, we cannot reasonably model the buyers as price takers. When several banks compete for a loan, the price is determined through bids submitted by the banks, i.e. in an auction. In the auction setting, the banks' rents do not arise from product differentiation in the usual sense. They are informational rents caused by the informational asymmetries between banks.

This auction approach to banking has been initiated by Broecker (1989). He models bank competition as a sealed bid common value auction where banks get an imperfect binary signal about the quality of the firm. The key insight of his paper is that if banks have independent imperfect information about the quality of a borrower, the screening activity of one bank has a negative impact on the other's profit. The firms refused by one bank will apply for credit at the other bank and reduce the average quality of this bank's loan applicants. In a model with binary signals, the competition of banks in interest rates has no pure strategy equilibrium. All firms that have been accepted by both banks will take the credit with the lower interest rate. Only firms, which have been refused by the bank quoting low interest rates, will go to the bank offering the higher interest rate. The average quality of these firms will be so bad that, despite the higher interest rate, this bank will make lower profits. It is therefore always profitable to slightly underbid the other bank and receive the good borrowers.

In Broecker's paper, the quality of the signal received by the firms is completely exogenous. A step further in the direction of endogenizing the ex ante information acquisition by banks has been made by Riordan (1993). He also models bank competition a sealed bid common value auction, but, in contrast to Broecker's model, banks receive a continuous signal of the borrower's quality. They are therefore able to quote a different interest rate for each borrower and the resulting equilibrium is in pure strategies. Riordan analyzes how the interest rates and loan approval decisions of each bank depend on the number of competitors and how the number of loans approved by the market changes with the market structure. However, he doesn't endogenize the effort deployed by banks to acquire information. The informativeness of the creditworthiness tests is still exogenous, Riordan only determines endogenously how the bank uses this information. Kannianen and Stenbacka (1998) try to completely endogenize the screening effort and analyze the socially inefficient distortions of the banks screening effort caused by competition. In a similar setting, Gehrig shows that if information acquisition is taken into account, monopolistic outcomes are stable with respect to competition. Unfortunately, both papers suffer from technical flaws. Indeed, endogenizing the information acquisition of bidders is still an unsolved problem in general auction theory. Matthews (1985) has formulated an auction with bidders of asymmetric information and endogenized the information acquisition in the context of pure common values and closed bidding. Despite assuming a quite simplified information acquisition technology, he is not able to ensure the existence of equilibrium. One way to get around the mathematical difficulties is to assume a discrete information acquisition technology and solve the mixed equilibrium strategies. This is how Hausch and Li (1993) proceed. Ruckes (1998) has adapted this method for modeling banking competition. He is able to show that competing banks will change their credit standards with the quality of the firms in the market. If there are a lot of good firms they will be much more lenient than if the risk of credit losses is high. Troege (2000) uses a similar model to show that bank ownership of shares may reduce credit market competition.

3.3 Monitoring and the lock-in effect

A number of papers have argued convincingly that free-riding effects may erode the banks' incentives to make firm specific investments in a highly competitive environment. For example Hellmann, Murdock and Stiglitz (1995a, 1995b) argue that too intense competition may lead to an underinvestment in deposit mobilization, inefficient allocation of resources and insufficient incentives to monitor firms. They argue in favor of entry restriction and (temporary) deposit rate regulation to reduce competition. Petersen and Rajan (1994,1995) show theoretically as well as empirically that less competition may increase the monitoring efforts of banks.

The opposite causality, i.e. the effect of monitoring on competition has also been examined in the literature: The idea that information acquired during a lending relationship may prevent competition has been pioneered by Sharpe (1990). He constructs a multi-period model of a lending relationship where, during the first period, the banks gain inside information about the quality of the firm. It is assumed that this information gives the bank an informational advantage, which prevents other banks to enter in competition at a later stage.

Using again an auction model, Fischer (1990) studied the competition of an informed and an uninformed bank in detail and pointed out that Sharpe was not entirely correct in assuming that privileged information about a borrower's quality prevents competition. In fact, the strategic situation will result in equilibrium with mixed strategies where insiders will earn higher rents, but outsiders will sometimes obtain the credit.

The same competition model was used in Rajan's (1992) analysis of the consequences of ex post rent-extraction. He compares the incentives for the firm provided by informed versus arm's length finance. Bank finance reduces the possibility of the entrepreneur to increase the risk of his project, but also reduces the entrepreneur's incentives to invest effort, because he anticipates rent extraction. Rajan concludes that borrowers who anticipate a sequence of very profitable projects would prefer arm's length or market financing to bank finance. Both Rajan (1992) and Sharpe (1990) assume that inside banks can extract rents, because they have better information about the borrower's credit risk. However, Troege (2000, 2001b) shows that monopoly power gained by information about the quality is relatively weak. He argues that it is not information about the quality but information which allows the bank to reduce credit risk which prevents competition.

3.4 Reputation

Bank firm relationships stretch over a long time horizon. This means that confidence building and reputation effects will play an important role to reduce asymmetric information and moral hazard on the borrower as well as on the lender side.

The impact of the debtor's reputation on his cost of finance and project choice has been analyzed by Diamond (1989) and (1991). The reputation building of bank is modeled in Sharpe (1990), Cemmanur and Fulghieri (1994) and Dinc

(1997) have pointed out the importance of the bank's reputation. Sharpe (1990) assumes that inside banks have therefore the possibility to extract ex post monopolistic rents and demonstrates how in a simple infinitely repeated game, banks can acquire reputation not to exploit their informational advantage at the borrower's expense. Cemmanur and Fulghieri (1994) focus on the decision of the financier to liquidate a firm in financial distress or renegotiate the credit. They show that reputational concerns will incite banks to make more costly evaluations of the firm than bondholders in order to take the correct decision. Dinc (1997) considers the incentives of banks to observe discretionary loan commitments. Troege (2001) demonstrates that in a repeated game banks will treat customers correctly in order not to lose profits from future financing relationships. However, the main focus is not the reputation mechanism itself but its possible disadvantages. It is demonstrated how the necessity to maintain a good reputation may lead to credit rationing.

4 Implications for Bank Management

A better understanding of the role of bank has important implications for bank management. The lessons from the microeconomic analysis of bank firm relationships can be summarized in two main points: 1) banks are fundamentally different from producing industry 2) bank finance cannot be understood in terms of market-based financing through bonds.

We argue that most management consultancies as well as banks themselves miss the implications of these insights for the management of banks. They understand the lending business in terms of simple producing industry and they analyze loans with statistical tools derived from bond portfolio theory.

The fact that banks are not in the first place a supplier of finance but a provider of information in the forms of screening and monitoring has several important implications. An immediate consequence is that the ability to collect and evaluate information and not the level of the interest rate alone determines the efficiency of banks. A bank which is more successful in selecting good firms will not only be able to make profits on these credits, it will also inflict losses on less informed competing banks, because these banks will provide credit to the firms refused by the first bank.

Similarly the ability to monitor firms closely will enable banks to act profitably in environments where banks practicing non-informed arm's length financing cannot lend. In most European countries cooperative banks and savings banks excel in this kind of relationship lending, although regulatory requirements my severely hamper the banks ability to lend to small non-rated firms. In the United States where strict regulation prevents banks from getting too closely involved in the management of firms, the role of monitoring lenders has been occupied with much success by venture capitalists.

Understanding that information acquired by the bank plays a crucial role also highlights the importance of loan officers and local bank branches. They should not only be considered as a costly distribution system of the bank's product but as the selection and supervision device which allows the bank to benefit from non-formalized information. Remuneration of credit managers should incite them to acquire and use information efficiently. Relying on computerized scoring and credit supervision techniques instead of the experience and judgement of credit managers will not only make the information less reliable, it will also homogenize the information owned by different banks and therefore lead to an erosion of rents. Last not least the reputation of bank is probably not attached to the organization as a whole but firms trust their local credit managers. Replacing them in order to pursue short-term profit oriented strategies may destroy confidence and have severe consequences for the long term performance of a bank.

Similarly identifying credit management with the management of a bond portfolios will probably destroy the competitive advantage of banks with respect to bond markets thereby pushing firms into desintermediated finance. A bank can sell a loan, however the bank's knowledge about the company can probably not be transferred to another bank. Therefore transferring loans will increase credit risk. Exploiting the possibility of banks to influence firms and reduce credit risk instead of simply trading credit portfolios, should help banks to gain competitive advantages not only compared to other banks but also with respect to the bond markets.

5 Policy Implications

The research reviewed above also has important consequences for the current discussion of prudential regulation. Again there are two major insights regarding regulatory policy: 1) the competitive advantage of banks is the utilization of non formalized information and 2) banks can actively influence the riskyness of a credit. It is evident that risk control systems which are relying on formalized historical data for evaluation the riskyness of a given credit portfolio cannot take that into account and will therefore lead to a distortion of the banks' investment strategy.

A number of economists have provided evidence that these distortions in the allocation of capital can have important consequences for the evolution of an economy. Mayer (1988) has pointed out that banks are the most important source of external finance in all main capitalist industrialized nations. In a more recent paper Carlin and Mayer (1998) collect evidence that different financial systems not only influence overall growth, but are suited to different types of industrial activities and different stages of developments of firms and economies.

More detailed empirical evidence that structure and regulation of the banking sector influences the economy is provided for example by Jayaratne et al. (1996).

Their paper demonstrates that even the fine-tuning of the banking regulation can be decisive. They examine the liberalization of the banking sector in different states in the United States and show that this had a positive influence on a state's growth. Cetorelli and Gambera (2001) show that bank concentration facilitates credit access to younger firms but depresses overall economic growth.

Policy makers have raised similar concerns. For example the German Monopolkommission (1998) expresses strong concerns about the availability of finance for innovative firms due to the bank dominance of the financial system. Authors like Porter (1992) have for example pointed out the drawbacks of market-based systems. He argues that strong banking regulation favors the predominant stock market financing in the US and leads to a myopic concentration on short-term returns. All these observations suggest that small institutional or regulatory differences may have strong consequences. They may change the overall growth and direct the evolution of the economy in one or another development path.

But what are the alternatives to a control based approach to regulation? The classical advice from an economist is of course not to regulate through control but to implement incentive compatible efficient structures. How these structures would have to look like is unfortunately still a largely open question. A recent paper by Kose et al. (2000) has made first step in this direction by analyzing the impact of managerial compensation schemes on the banks risk policy. However, clearly more research on this topic is needed. All we can say at the moment is that there is probably no optimal "one fits all" type of banking regulation. Depending on market concentration, the importance of desintermediated markets, the structure of the producing industry and most importantly the extend to which banks engage in relationship or in arm's length finance optimal regulatory systems would have to look very different.

References

Allen, F. A. and D. Gale (1995), A Welfare Comparison of Intermediaries and Financial Markets in Germany and the US, European Economic Review, 39:179-209.

Amable, B., J. B. Chatelain and O. DeBandt (1997), Stability versus Efficiency of the Banking Sector and Economic Growth, Banque de France , Mimeo.

Aoki, M. and S. Dinc (1997), Relational Financing as an Institution and its Viability under Competition, Stanford University, Mimeo.

Arrow, K. (1964), The Role of Securities in the Optimal Allocation of Risk Bearing, Review of Economic Studies , 32(2):91-96.

Boot, A. W. and S. I. Greenbaum (1993), Reputation and Discretion in Financial Contracting, American Economic Review, 83, (5):1165-1183.

Broecker, T. (1990), Credit-Worthiness Tests and Interbank Competition, Econometrica, 58 :429-452.

Bundestag (1997), Protokoll der 74. Sitzung des Rechtsausschusses, Rechtsausschuss , Deutscher Bundestag, Bonn.

Caminal, R. and C. Matutes (1997), Bank Solvency, Market Structure and Monitoring Incentives, Barcelona, Institut d'Analisi Economica, Mimeo.

Caminal, R. and C. Matutes (1997) Can Competition in the Credit market be Excessive? CEPR Discussion Paper No. 1725.

Carlin, W. and C. Mayer (1998), Finance, Investment and Growth, Oxford , mimeo.

Cetorelli, N. and M. Gambera (2001), Banking Market Structure, Financial Dependence and Growth: International Evidence from Industry Data. Journal of Finance, forthcoming.

Chan, Y.-S., S. Greenbaum and A. V. Thakor (1986), Information Reusability, Competition and Bank Asset Quality, Journal of Banking and Finance, 10:243-253.

Chemmanur, T. J. and P. Fulghieri (1994), Investment Bank Reputation, Information Production and Financial Intermediation, Journal of Finance, 49:57-79.

Chiappor, P. A., Perez-Castillo and F. Verdier (1995), Spatial Competition in the Banking System, Localization, Cross-Subsidies and the Regulation of Interest Rates, European Economic Review, 39, (5):889-919.

Cordella, T. and E. Levy Yeyati (1998), Financial Opening, Deposit Insurance and Risk in a Model of Banking Competition, CEPR Discussion Paper No. 1939.

Debreu, G. (1959), The Theory of Value, New York, Wiley.

Diamond, D. (1984), Financial Intermediation and Delegated Monitoring, Review of Economic Studies, 40:393-414.

Diamond, D. (1989), Reputation Acquisition in Debt Markets, Journal of Political Economy, 97:828-862.

Diamond, D. (1991), Monitoring and Reputation: The choice between bank loans and directly placed debt, Journal of Political Economy, 99:688-721.

Dinc, S. (1997), Bank Reputation, Bank Commitment and the Effects of Competition in Credit Markets, Mimeo, University of Tokio.

Economist (1998), American Financial Regulation: The Great Escape, The Economist, March14th -20th: 98.

Fischer, K. (1990), Hausbankbeziehungen als Instrument der Bindung zwischen Banken und Unternehmen. Eine theoretische und empirische Analyse, Univ. Diss, Universitaet Bonn.

Freixas, X. and J.-C. Rochet (1997), Microeconomics of Banking, MIT Press, Cambridge, Massachusetts.

Garella, P.G. (1996), Informational Rents in Interbank Competition, Recherches Economiques de Louvain, 62, (1):3-20.

Gehrig, T. (1996), Market Structure, Monitoring and Capital Adequacy Regulation, Swiss Journal of Economics and Statistics, 132, (4/2):685-702.

Gehrig, T. (1998), Screening, Cross-Border Banking and the Allocation of Credit, Universität Freiburg, Mimeo.

Goldsmith, R. (1969), Financial Structure and Developement, Yale University Press, New Haven.

Gorton, G. and A. Winton (1998), Banking in Transition Economies: Does Efficiency Require Instability? Journal of Money Credit and Banking, 30:621-655.

Hausch, D. and L. Li (1993), A Common Value Auction Model with Endogenous Entry and Information Acquisition, Economic Theory, 3:315-334.

Holmstrom, B. and J. Tirole (1997) Financial Intermediation, Loanable Funds and the Real Estate Sector, The Quarterly Journal of Economics, 112, (3):663-691.

Jayaratne, J. and P. E. Strahan (1996), The Finance Growth Nexus: Evidence from Bank Branch Deregulation, The Quarterly Journal of Economics , 111:639-670.

Kanniainen, V. und R. Stenbacka (1998), Project Monitoring and Banking Competition under Adverse Selection, Wissenschaftszentrum Berlin fuer Sozialforschung, Discussion Paper 97,23.

King, R. and R. Levine (1993), Finance and growth: Schumpeter might be right, The quarterly journal of economics 108 :717 - 737.

Klein, M. (1971), A Theory of the Banking Firm, Journal of Money Credit and Banking, 3 :205-18.

Kroszner R. S. and P.E. Strahan (1999), Bankers on Boards: Monitoring, Conflicts of Interest, and Lender Liability, NBER Working Paper No. W7319.

Mondigliani, F. and M. Miller (1958), The Cost of Capital, Corporation Finance and the Theory of Investment, American Economic Review 68:261-297.

Monti, M. (1972), Deposit, Credit and Interest Rate Determination under Alternative Bank Objectives, In: Mathematical Methods in Investment and Finance, edited by G.P. Szego and K. Shell, Amsterdam North-Holland.

Monopolkommission (1998), Ordnungspolitische Leitlinien für ein funktionsfähiges Finanzsystem, Sondergutachten 26, Nomos Verlagsgesellschaft, Baden-Baden.

Petersen, M. and R. Rajan (1995), The Effect of Credit Market Competition on Lending Relationships, The Quarterly Journal of Economics, 110 , (2):407-443.

Petersen, M. and R. Rajan (1994), The Benefits of Lending Relationships: Evidence from Small Business Data, Journal of Finance 49:3-37.

Rajan, G. (1992), Insiders and Outsiders: The Choice Between Informed and Arms-length debt, Journal of Finance, 45:1367-1400.

Rajan, G. (1995a), Convenants and Collateral as Incentives to Monitor, Journal of Finance, 50:1115-1147.

Rajan, R. (1995b), The Entry of Commercial Banks into the Securities Business: A Selective Survey of Theories and Evidence, University of Chicago, Mimeo.

Riordan, M.H. (1993), Competition and Bank Performance: A theoretical Perspective, In: C.Mayer and X. Vives, editors, Capital Markets and Financial Intermediation, Cambridge University Press: 328-343.

Ruckes, M. (1998), Bank Competition, Credit Standards and Corporate Conservatism, Working Paper Universität Mannheim.

Kose, J. , Saunders, A. and L. W. Senbet, (2000) A Theory of Bank Regulation and Management Compensation, Review of Financial Studies 13 (1)95-125

Salop, S. (1979), Monopolistic Competition with Outside Goods, Bell Journal of Economics, 10, (1):141-156.

Sharpe, S. (1990), Asymmetric Information, Bank Lending and Implicit Contracts: A Stylized Model of Customer Relationships, Journal of Finance, 45:1069-1087.

Kannianinen, V. and R. Stenbacka (1998), Project Monitoring in Lending Markets with Adverse Selection, Swedish School of Economics, Helsinki, Mimeo.

Troege, M. (2001), "Should Banks be Allowed to own Equity in Non-financial Firms?", ESCP-EAP, Paris, mimeo.

Troege, M. (2000), "The Insider's Curse", ESCP-EAP, Paris, mimeo.

Troege, M. (2001), "Monitored Finance, Usury and Credit Rationing", ESCP-EAP, Paris, mimeo.

Yanelle, M. O. (1997), Banking Competition and Market Efficiency, Review of Economic Studies, 64:215-239.

Yanelle, M. O. (1989), The Strategic Analysis of Intermediation, European Economic Review, 33:57-84.

From Banking on Customers to Banking on Regions: A Flexible, Diffusive Bricks & Mortar Network Organization for Real Focus on Customers

D.G. Karellas , D.S. Stamoulis

Alpha Bank
Stadiou 40
105 52 Athens, Greece

Summary:
The organization of the main delivery channel of banks, the branch network, is of critical importance for the overall bank's success. The evolution of a bank's branch network organizational models is discussed and the paradigm shifts from banking on products to banking on customer and then to banking on regions / communities is explained, together with their implications onto various functions of the bank. The most commonly used community banking model cannot address two problems – local clientele of different customer segments and exploitation of small-scale business opportunities – and, thus, the extended community banking model is proposed.

Keywords: Community banking, Branch network re-organization, Branch redesign, Customer contact

1 Introduction

Traditionally, bank branches have been almost autonomous, self-sufficient banks, by offering products and processing client requests internally, with little interaction with the rest of the bank's divisions. Highly product oriented rather than customer focused, branches have been fulfilling customer orders instead of actively promoting and selling products / services to customers. The rationale for competing on geographical proximity with potential customers led the banks to constantly increase the number of their branches. The density of a branch network

---

The authors express personal ideas of their research effort, and, therefore, the ideas presented here do not necessarily reflect, represent or coincide with those of the Bank.

had been synonymous to a bank's growth. However, this expansion cannot go forever, on the one hand. On the other, no matter how dense a branch network may be, not all business opportunities for banks can be exploited, since it is often the case that the operating costs for a bank branch exceed the benefit of exploiting an emerging or seasonal or limited scope banking business opportunity in a particular region/ community.

By explaining how to establish different types of branches and organize them according to the needs of the region, this paper proposes a new branch network organization model, redefining the concept of community banking. The aim is to show how banking business opportunities can be exploited in a cost-effective way, without recourse to an endless expansion of the branch network.

In this paper, the organizational model paradigms of a bank's branch network are discussed, and a new, extended, community banking model is proposed as a solution to a two-fold problem: (a) how to best focus banking units spread within a region / community to the specific needs of the local micro-markets, and (b) how to cover a whole region / community without opening new branches everywhere, including some banking business opportunities that are small and unstable enough for a bank to maintain a branch all over the year. Besides the fact that a typical branch does not exist as explained in this paper, the proposed extension of the community banking model as proposed, offers a solution to the following problem: *how a bank should diffuse its activities in order to be able to reap the benefits of emerging, seasonal or permanent economic activity that is not big enough to justify the opening of a new branch, but, this opportunity is, nevertheless, potentially profitable for financial intermediation, which is the core banking business.*

In parallel with the network organization, we observe how the internal branch organization evolves, since the entire banking strategy regarding the distribution channels is reflected not only onto the level of the branch network organization, but also onto the internal branch organization philosophy. The implications of those strategic changes are far reaching and affect other functions of a bank such as accounting, MIS, information systems infrastructure, procedures, job types etc. Examining the internal as well as network organization of the branches, a table has been drafted to summarize the core ideas that describe the paradigm shift from product- to customer- orientation and then to regional/community-orientation.

Section 2 discusses the drawbacks of the old branch network organization models. Section 3 introduces the extended community banking model as a way to overcome the limitations of the tradtitional community banking model. Section 4 epitomizes the philosophy of the new model as a paradigm shift from banking on customers to banking on regions and section 5 presents concluding remarks.

2 Reengineering the Old Model for Branch Network Organization

From the organizational point of view, banks used to be a corporate headquarters for central single functions such as the dealing room, technology, marketing, central financial management etc. plus a number of autonomous branches, which were small banks per se. An important indicator of a bank's growth and success was the number of its branches, since a bank's market share was strongly related to its network size, in their earlier expansion phases [Ghosh, 1987].

However, the opening and running of new branches is associated with high costs, not only because the branch has to be established in a high visible and accessible area meaning high premises costs, but also because of labour and operating resource costs. Therefore, "the larger institutions are now actually aiming to reduce the overall size of their networks while maintaining and even increasing their market shares" [Greenland, 1994].

"Dramatic changes in branch banking will manifest themselves not so much in the total number of branches as in what those branches look like and what happens in them day to day" [Mendonca, 1996]. The main, obvious restructuring decision for branches is, of course, to move the back-office / administrative type of work away from the branch. This decision is not only dictated by the need to optimise the branch for face-to-face banking but also by the economies the scale. Specialization, accuracy, consistency and speed should characterize back-office operations, which does not require the physical presence of the customer. However one cannot expect to reap these fruits unless this type of work is centralized. What remains left in the branch must only be sales and service. "One outcome of centralized processing has been for banks' manufacturing activities to be physically separated from their selling activities. Previously, these functions had co-existed within the branch. The modern bank branch constitute the bank's salesroom and has been extensively revamped to fulfill this objective" [Orton, 1995].

Organizing the network along the front- and back-office separation lines, branches loose their autonomy and become dependent to branch support centres. The branches focus on identifying customer needs, perform financial planning for their customers, provide advice, initiate the sales process, cross-sell and up-sell products and customers [1]. The role of a branch's staff is to offer financial solutions tailored to the needs of an individual, a household or a business.. Refurbishing the branch for those tasks, customer relationship management is leveraged for maximum effectiveness. The branch support centre analyzes the data, prepares the decision making process, arranges for payments, caters for the accounting settlement of the customers and the branches, controls and ensures the quality of

---

1 "Branches will no longer be managed by bankers but salespeople who know banking" [Svigals, 1996]

data that is input into the bank's information systems, detects potential fraud situations etc.

3 The New Model: Community Banking Revisited

3.1 Branch Network Hierarchy

Identifying the type of prospect clients in a micro-market in conjunction with a bank's strategy as far as its presence in a region/community is concerned (what the bank is aiming to do in each micro-market), a branch configuration that is most suitable for each micro-market is determined. Therefore, the branch network is a mixture of retail outlets for retail/consumer customers only, commercial banking branches for small and medium size corporations, corporate banking centers etc. Obviously, the retail outlets are numerous and geographically dispersed; the commercial banking branches are fewer and even less in numbers usually the corporate banking centers.

"When branch networks of the larger institutions are observed as a whole it is clear that different types of service outlet and administrative centres have evolved. When these are viewed according to the different levels or order of service than they provide, a 'hierarchy' of representation can be identified" [Greenland, 1994]. Thus, the financial services industry network hierarchy consists of "remote ATM and/or deposit box, automated branch, sub-branch (small-retail unit), parent branch (retail/personal branch), corporate branch, community branch / financial supermarket, regional headquarters/ administration and control centres, national headquarters" [ibid]. Consequently, at the heart of a branch network strategy lies the rationale for selecting the appropriate element of the branch network hierarchy for representing the bank in each geographic region / community.

Before opening a new branch, qualitative and quantitative market research reveals the type of economic activity that would be profitable for a potential banking business and, thus, the relevant type of branch targets the corresponding customer segment. In a residential area, a small retail outlet or a retail/personal branch is enough. In an area full of corporations, the bank is present with a corporate banking business centre. Needless to say, each element of the branch network hierarchy is endowed with such skilled staff, expertise and resources that are necessary to address the needs of the type of clientele aimed to serve.

"Primarily in a attempt to make the branch network more cost effective, individual branches no longer necessarily duplicate each other's functions by providing a full range of services. Instead branch networks are arranged in clusters centered around a 'key branch' which offers a full range of services. The peripheral or «satellite» branches provide a restrictive service, concentrating either on the personal or corporate banking sectors" [Howcroft, 1996].

3.2 Community Banking Model

Banks are applying this model in the recent years with a view to adapt the mode of their presence to the higher benefit-lower risk customer segments on the one hand, and, on the other, to concentrate their power on sales and service cost-effectively in each of the micro-markets targeted. Deciding which element of the branch network hierarchy to be used is actually a cost-benefit analysis, and, more precisely, such a decision is based on a investment plan for each region/community where a bank operates. The term community banking has been coined to describe this model, because of the tailoring of the banking presence – both organizational units and product offering – according to the anticipated type of banking business in the local area / micro-market / community. However, it is not often the case that regions / communities are populated by a singly type of customer segments. In some countries, such as in the UK, the industrial zones, the office building areas and the residential areas are clearly distinct in some regions. In these cases the application of the community banking concept as used up to now is relatively straightforward, since a branch is either retail- or commercial-oriented. But even if a bank identifies a residential area as a potential micro-market, the type of residents in region/ community rarely belongs clearly and solely to one customer segment only. Most probably, the residents will belong to the mass-market segment, but some of them may belong to the affluent segment, or even to the very high net worth segment. All three of these customer segments require differentiation in products, in levels of service and in advice. Therefore, the community banking model – meaning that the appropriate element of the branch network hierarchy is selected for the each region / community – should be extended to cover the different requirements within the targeted region / community.

3.3 The Open-Platform Branch

Instead of establishing separate branches to serve the various customer segments existing in a region / community, the extended community banking model proposed here advocates that there exists only one branch type, that of the open platform one. The new extended model views the elements of a branch network hierarchy as modules of the open platform branch, instead of treating them as distinct organizational units. The main idea of the extended community banking which solves the problem of the various customer segments within a region / community is that the branch consists of modules / components that are destined to serve differentiated customer needs. Being an open platform, each branch is composed of any combination of building blocks – i.e. the aforementioned modules / components – according to the specific needs of its micro-market.

The open platform branch is usually made up of a combination of any of the following modules/ building blocks:

- the branch management area,
- the cash transactions area (for coins, banknotes, cheques and foreign exchange)
- the retail banking module,
- the commercial banking module,
- the corporate banking module,
- the private banking module,
- the special customers' module, e.g. shipping,
- a private interview area,
- the automated lobby / self-service electronic banking area.

Practically speaking, the componentization of the branch allows customers to be served in separate modules of the same open platform branch by different staff members and with different levels of service quality. For example, a branch may comprise of a retail and a commercial banking module, another one in a residential area may have a retail and a private banking module. These modules need not necessarily share the same entrance, nor operate concurrently during the traditional banking hours. Being separate modules, branch building blocks may operate autonomously, but they can also take profit of their co-location, since a customer of one module may be an ideal sales hint for an another module. Therefore, modules can leverage each other's sales, as the following example shows: a high net worth man enters into the suitably decorated private banking area but his household accounts may only be served by the retail banking module.

Since branches «traditionally, [they] relied on the provision of basic encashment facilities to attract customers but as this incentive to enter branches recedes in the wake of alternative technology-driven distribution channels, the problem of how best to attract customers into branches will become of paramount importance» [Howcroft, 1996]. By co-locating the modules destined to serve different customer segments, this potential threat is averted as the different open platform branch modules create synergies among them for an optimal management of the customer relationships.

For all those reasons, the flexible, modular configuration of the open platform branch allows for maximum adaptation of the branch network distribution channel to the local needs of the region / community in which it is located. The particular configuration of this branch is a result of a situational analysis aiming at maximizing the efficiency and effectiveness of every new branch.

Besides those advantages, the extended community banking model has to address one more issue. That of exploiting small scale opportunities arising into a micro-market. Suppose a bank has already established an open platform branch in a region. Within this region / community, the need for financial services is not always normally distributed around some spatial "concentration" points. The demographics of the local clientele, the diversity of economic capacity of the potential clientele segments, the emerging business opportunities, the buying patterns, some seasonal activity create differentiated needs for tailored banking presence. What type of organization is most appropriate in order to cover a whole region that exhibits periodically or constantly some economic activity that is not big enough to justify the opening of a new branch?

3.4 Provisions by the Extended Community Banking Model

Thus, the extended community banking model has to propose a new method for exploiting banking business opportunities that are
- either outside the reach of the operational range of their branch network, or
- of an economic activity that is not big enough to justify the cost of a branch, or
- unstable due to their nature (seasonal, emerging, high risk etc.), or
- any combination of the above.

In all these cases, the old bank organization thinking would dictate the opening of more branches. Such a strategy is meaningful only when a region/ community is heavily underbanked and, therefore, the density of the branch networks in this region / community allows the opening of new branches. Otherwise, the reaping of the benefits from such business opportunities is being annihilated but the cost of maintaining a branch.

If an opportunity of the above type is identified but the economic activity is outside the reach of the operational range of their branch network, establishing more branches would create problems:

1. Within a wide region/ community, several seasonal, emerging, or special type business opportunities may exist; these opportunities require the offer of specialized products and not the whole range of the banking and financial products and services. Thus, reserving much more than necessary of the physical and human resources that are usually allocated to a bank branch has no business sense, and obviously does not justify the investment.
2. In a small-scale market, a bank need not get involved in every type of banking and financial activity. The rule of thumb: '40% of the economic activity creates more than 60% of this market's profit' gives an indication of how a bank should penetrate a small-scale market. Identifying the high net worth and low risk financial market share per customer segment and economic activity, long-term profitability through customer relationship management can be achieved. Furthermore, profit can accrue when a specific banking service is repetitively offered within an appropriate for this purpose organizational setting. Typical case in this category is the exchange bureau. If we try to establish a full branch instead of an exchange bureau in an area where the main activity of banking interest is the foreign exchange, certainly the branch will not be successful. Generally speaking, if an organizational unit of a bank tries to capture indiscretionally as much of the local micro-market as possible, the portfolio of this branch will severely deteriorate. The quest for market share inevitably leads to the degradation of the results and the situation is that the main revenue generating products & services try to balance the costs incurred by such an unneeded costly presence.

3. When a new opportunity for banking/ financial products and services appears and lies outside the operational ranges of the nearest branches, it sometimes arises a counterproductive competition among those branches.
4. The organizational set-up of a complete branch (administration, accounting, staffing, room etc) is too 'heavy' for a unit that has to be very flexible to recognize, penetrate and react to the discontinuities of the normal distribution of the potentially profitable economic activity not close enough to it.
5. Low exploitation of the technological capabilities which allow the customer to choose the time and place of examining and deciding upon the bank's offerings.

To explain what the solution to this problem can be let's consider a scenario. A branch has been established in the capital of an island to serve the local needs. The branch has been composed of the appropriate modules focusing on the basic customer segments identified in the capital. In the port, which is some kilometers away from the capital, people need to spend small amounts of money before and after the embarkment and during waiting. Therefore, an off-site ATM is put there for this purpose. This is in essence a detachment of the cash transactions services of the branch in the capital placed at the port. About 40km south of the capital, beautiful beaches are full tourists who stay in the near-by hotels for their vacations. A foreign exchange bureau offers then on-site service, to turn this seasonal economic activity into profits for the bank and the branch at the capital, more specifically. However, there are hardly any tourists during the winter and the people who live in this area are interested in investing the money earned during summer. Furthermore, the natives are also dealing with agriculture, and, thus, the summer operating foreign exchange bureau is transformed during winter into a small investment advice office and leasing service desk for agricultural equipment. Moreover, plenty of rent-a-car businesses are located 60km north of the capital because visitors rent a car to make a small tour of the mountain that is a physical shelter for endangered species of birds and butterflies. The 'long-arm' of the capital's branch has also been established there as a unit for on-the-spot selling of car and driving insurance. In the east of the island, a higher education institution campus is open from October to end of May and is closed in summer. Students and staff can use an ATM for cash and may ask for a consumer loan, for example, at a small retail outlet within the main building of the campus staffed with two retail banking officers. During summer, these two persons return to the branch in the capital of the island to back-up other colleagues who are absent for holidays.

The imaginary scenario explained above speaks eloquently of how banking units can diffuse their mode of presence within a region / community in order to exploit opportunities otherwise unreachable by a branch network of a given density, whenever the cost of opening a new branch cannot be justified. But, nevertheless, these flexible units offer targeted banking services as if they were modules within the open platform branch. By adapting the type of presence as well as the offerings according to the nature and the characteristics of small-scale banking business opportunities, redefines the community concept in way that decouples it both from the spatial constraint of the distance between a branch and a customer and from the static identification of customer segments which populate the 'catchment' area around a branch. Needless to say, competition has also to be taken into account when decisions about the type of banking units within a region are being made. It is outside the scope of this paper to analyze the impact of existing competitive banking units withint the same region or 'catchment' area of a branch. The proposed model does not ignore competition, but has a different focus.

3.5 Managing the New Region / Community Banking Model

"Even though Asset Liability Management systems have been built, and profit-oriented types of control indices have been introduced in order to improve the profit- and risk-control systems of the financial industry, one of the root causes of their unsatisfactory effects is that *profit centres have not been endowed with enough flexibility to contend with changes in business environments and structures* {emphasis added}. There are basically two dimensions – area and service – in establishing a profit centre. Its optimum structure obviously varies depending on the type of business strategy adopted by each financial institution. At present, all financial institutions establish their profit centres *by branches and by area* {emphasis added}. These profit centres function quite effectively for financial institutions that limit their operational territory and handle only basic products and services" [2]. That is exactly the point made by the extended region/community banking model. Our approach advocates that profits can only be optimized when a flexible set of self-transforming banking units is spread all over the place of a region / community to address specific fruitful economic activity. A web of banking units that may expand or shrink like a bubble is replacing the old idea of the branch network, implying that the cost and profit center is obviously this web of units as a whole and not each one of the individual units. Financial management, accounting issues and the like are catered for by the core of the web, which is an open platform branch. The relevant modules of the open platform branch operate as a 1st level support center for those web units that serve the same customer segment. Since the banking units of this web organization may change the product offering mix from time to time and from place to place as explained in the scenario of the previous section, it is meaningless to measure independently each one of those units.

In general, the allocation of resources in the extended region/ community banking model is not a one-time decision, but rather a continuous process of adaptation to the local and contemporary economic activity. The lack of such

---

2 Excerpt from Chapter 3.4, *The enhancement of Corporate Vitality by the Innovation of Management Systems*, pp. 108-109, Yamaguchi et al. (1991)

flexibility in defining the mode of presence, the product mix and the appropriate skills prohibits a bank from responding innovatively to a changing environment, and, thus, indicates that "organizational structure is out of alignment with organization needs" [Daft, 1995], which has been identified as a key symptom of organizational deficiency [ibid].

4. From Banking-on-Products to -Customers and to -Regions

The following tables summarize the paradigm shift observed so far and anticipated in the near future, regarding the organization of the branch network and underpins the implications of each model on various areas of a bank's organization.

MAIN ORIENTATION: *Banking on ...*			
	PRODUCT	CUSTOMER	REGION
FOCUS ON	Financial need	Need of a person or of an enterprise	-Same types of needs of persons / enterprises -Needs of same type of persons / enterprises
BUSINESS DEVELOPMENT	Mass production of products	Mass distribution seeking for Customer Loyalty *[based on the rule "3 products with us- no customer defect"]*	Exploitation of opportunities stemming from economic activities whenever and wherever these appear
TARGETS	Distribution of as many products as possible	Acquisition / retention of high profit – low risk customers	Flexible response to regional opportunities with optimal type of presence
PROFITS	By sales High volumes of products	Profitable relationships with customers: -for a long time -3 products per customer.	Milking of healthy economic activity in the region

MAIN ORIENTATION: ***Banking on …***	PRODUCT	CUSTOMER	REGION
CORE CAPABILITIES	Rapid increase of -sold product volumes -market share per product	-Retention / expansion of market shares -To get the most out of each customer	Financial intermediation into any potentially profitable economic activity within region [*in any point, at any time, for as long as remains profitable*}
FIELD OF COMPETITION	-The earliest possible identification of needs to be covered -Engineering of new products	-Targeting customer segments -Quality in managing the relationships with customers	Any healthy economic activity all-year / seasonal, appearing at any place
CIRITICAL SUCCESS FACTOR	Wide range of products	Customer Relationship Management	Organizational flexibility for sales & service, of base-unit & sub-units covering the region
MARKETING TARGETS	-Elicit requirements -Understand needs -Design appropriate product	Sales hints to identify opportunities per existing customers and prospects	Identify areas within a region potentially constituting a banking opportunity
SALES TARGETS	To satisfy an expressed or latent customer need	-Cross-selling -Uup-selling -Value addition to customer relationships – Maximization of customer's potential for profitability	Delivery of most profitable products for the given space-and-time coordinates in this region
MEASUREMEN FOCUS	Per product -revenue - cost -profit - risk	Per customer -revenue - cost -profit - risk	Per region / community, for the whole web of units that comprise a bank's presence in this region

MAIN ORIENTATION: *Banking on …*	PRODUCT	CUSTOMER	REGION
IMPACT ON ACCOUNTING & MIS	*Low-level information:* Accounts per product *Main aggregation level:* Branch	*Low-level information:* Accounts per customer *Main aggregation level:* Overall customer relationship with the bank	*Low-level information:* Overall customer relationship with the bank *Main aggregation level:* Overall banking activity in the region /community
IMPACT ON ORGANIZATIONAL PHILOSOPHY	Organization based on Product Grouping Bank's internal divisions Per product Branch internal divisions Per product	*Organization based on* *Customer Segments* *Bank's internal divisions Customer segments (*)* *Branch internal divisions Customer segments* *(*) and possibly per Group Product Managers*	Organization based on Regions/ Communities - Each community autonomously managed (for sales & service) - Optimal number of organizational units, tailored to the type of economic activity of the region /community - The region/ community is a single cost & profit center as opposed to the old model where each unit (e.g. branch) is a cost and profit center.

5 Concluding Remarks

As it is the case with products, delivery channel organization follows as well the same well-known S-type maturity curve. Classical business process reengineering (BPR) theory advocates that when a product or service or delivery channel has reach the maturity point at the top of the S curve, there is no point in expecting any other significant improvement of its performance, despite the amount of effort. A major leap forward can then only be achieved if a new thinking is applied to give birth to another maturity curve with higher effectiveness levels.

Using this idea, the following figure 1 depicts the history of the bank branching evolution:

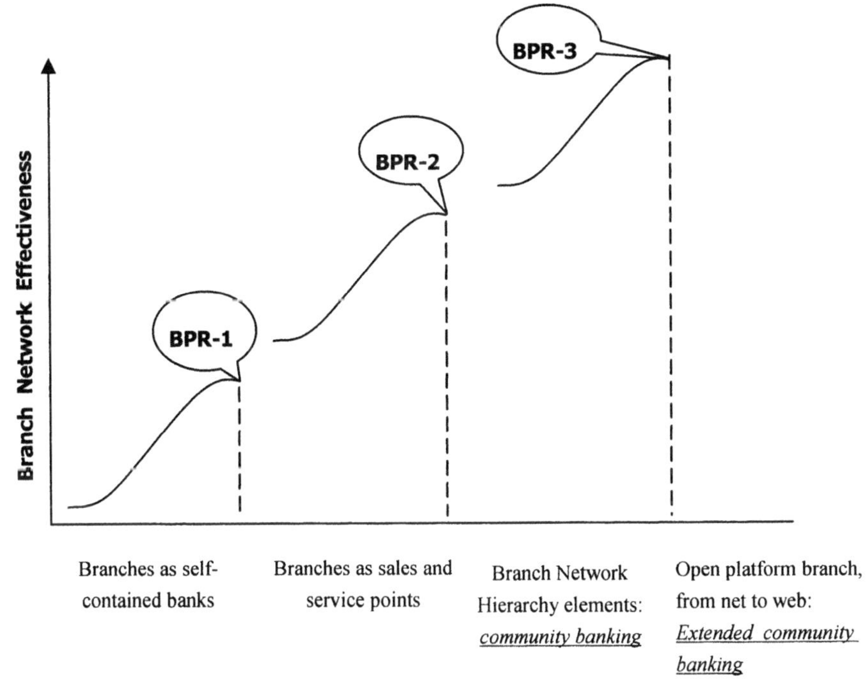

Figure1: the evolution of the branch network organization

This paper proposed an extension to the classical community banking model because the application of customer relationship management alone does not solve all the problems for a truly customer-centric organization. Banks tend to forget that customer relationship management (CRM) must be applied within a broader strategic framework, which the branch network organization model. Transforming the network into a web of banking units that surround an open platform branch with a view to reap the benefits of profitable but small-scale, unstable, seasonal etc. banking business opportunities, banks can create the appropriate context within which CRM can flourish. Componentizing the branch instead of selecting elements of a bank branch network hierarchy and then spreading a web of dedicated type of service banking units around it, a bank can diffuse its operation to a whole region / community, and, thus, significantly extending the 'catchment' areas of its branches.

The open platform branch circumvents the problem of servicing at the appropriate level of service and offering the correct range of products to the different customer segments who co-exist within the operational range of a branch. The economic rationale of the extended community banking model is simple. Human resources with the appropriate skills, premises and effort is allocated only when and where it is necessary, so that it makes financially sense to intermediate in a small-scale economic activity. Opening new branches all over the place cannot by justified by a cost-benefit analysis.

Further research on this topic will hopefully reveal best management practices for operating the somehow complex extended community banking model proposed here.

Acknowledgement

The authors thank Mr. John P. Parlitsis, Manager, for his contribution to the formation of some ideas presented in this paper and for his valuable comments.

References

Daft R. L. (1995) *Organizational Theory and Design.* West Publishing Company.

Ghosh A. and McLafferty S. L. (1987) *Location strategies for retail firms.* Lexington Books, Lexington, MA.

Greenland S.J. (1994) Rationalization and restructuring in the financial services sector. *International Journal of Bank Marketing*, 22(6), pp21-28

Greenland S.J. (1995) Network management and the branch distribution channel. International *Journal of Bank Marketing*, 13(4), pp12-18

Howcroft B., Beckett A. (1996) Branch networks and the retailing of high credence products. *International Journal of Bank Marketing*, 14(4), pp. 3-11

Mendonca L., Nakache P. (1996) Branch banking is not a dinosaur. *The McKinsey Quarterly*, No. 1, pp.137-146.

Orton I.(1995) *Reengineered Banks.* Lafferty Publications, Dublin.

Radecki L. J., Wenninger J., Orlow D. K. (1996) Bank Branches in Supermarkets. *Current Issues in Economics and Finance*. Vol.2, No.13, Federal Reserve Bank of New York.

Svigals J. (1996) *Bank Branching 2010*. Lafferty Publications, Dublin

Yamaguchi M., Morita Y., Dambara Y., Torii K., Halvorson E., Yamaguchi N., Matsumoto T., Yamashita K. (1991) *Banking Strategies for the 21st century*, Hitachi Research Institute, Tokyo. Printed in the USA.

Integrated Customer Relationship Management in the Banking Sector

M. Vlachopoulou[1], V. Manthou[1], O. Ioannidis[2]

[1] University of Macedonia
Department of Applied Informatics
Thessaloniki, Greece

[2] National Bank of Greece
Thessaloniki, Greece

Summary:
The shift from the enterprise-centric business model to a customer-centric business model is a challenge for the leading companies in the banking sector. Banks are trying to find new ways to understand, reach and satisfy their customers. The local point of this trend is CRM, a disciplined, holistic approach to identifying, attracting and retaining customers. CRM involves reengineering the organization's environment, from technology to processes to the skills and abilities of employees. The aim of this paper is to identify the goals of an integrated banking CRM system, to analyze the information needed from both internal and external sources as the necessary basis of data supply to a banking CRM system. A model of CRM process cycle in banks is developed, the appropriate CRM modules and key enabling technologies are defined, and the CRM application architecture is determined. Additionally specific data marts for banking applications are organized. Special focus is given to the advances made on the methodological tools that can be used to improve the operation, effectiveness and performance of the banking sector. Banks are linking and coordinating their efforts to develop new products/services fast and inexpensively and tailor them to the specific needs of different target markets or individual customers. By using CRM systems they are meeting customer expectations while at the same time achieving market share and profitability goals.

Keywords: Customer Relationship Management (CRM), Banking marketing, Banking Information systems

1 Introduction

The dramatic increase in competitive pressure for customer's loyalty, in modern global marketplace of financial services, pushes banks to redefine their overall business strategy, in order to align it with new market's requirements. Market dynamics that banks must address are the emergence of new technology, the increasing levels of customer expectations, the globalisation, the convergence of traditional markets and the new customer contact channels (Lawer 2000, Miller 2000).

Today's customers are more informed, more open to technology's innovations, more flexible in their dealing with financial institutes, and have more alternative channel and brand choices. As a result, they increasingly demand both upgraded banking products, tailored to their particular needs, and financial services with special emphasis on transaction convenience and banking functionality. The answer to this customer demands' pressure is the "one to one banking marketing" which focuses on each individual customer's financial needs, contact behavior and transactional history for all marketing activities (Vlachopoulou 1999).

Banking companies are increasingly adopting customer-centric strategies as a means of maintaining and enhancing their competitive advantages. Customer Relationship Management (CRM) is a concept that requires a new customer - centric business model which focuses on the seamless integration of every business that touches the customer. It provides bank institutions with a means to conduct interactive, personalized and relevant communications with customers across both electronic and traditional channels. CRM is now becoming the most advantageous way of realization an effective "one to one" marketing strategy, in order to attract, retain and expand profitable relationships with customers (Sheth and Parvatiyar, 1995).

An integrated banking CRM system supports a complete view of the customer, coordination of customer touch points, customer data analysis for personalized value proposition, marketing planning to increase response rates and effectiveness, and dynamic customer metrics.

The above issues, define the requirements of an integrated customer relationship management system in the banking sector and the consequent changes in the organization's culture, systems, processes, management and marketing activities.

In this paper a model of CRM process cycle in banks is developed, depicting the complete framework for an integrated CRM. For every cycle phase the specific processes, activities and modules, as well as enabling technologies are determined. Furthermore, the appropriate CRM application architecture is designed, and specific data marts for banking applications are organized.

2 The Main Goals of an Integrated Banking CRM

The main goals of an effective Integrated CRM solution in the banking sector, are to enable financial institutes to: a) widen customer relationship through acquiring new customers, identifying and targeting new segments and expanding in new markets, b) lengthen the existing relationship developing longer term relationships, increasing perceived value of products, and introducing new products and c) deepen the relationship with customers initiating the cross selling and up selling opportunities, understanding the propensity of different customer segments to purchase and increase sales (Peter 1998, Chablo 1999).

The implementation of a CRM system in a bank helps the business organization to obtain a complete picture of their existing customers, design both customer-oriented and market-driven financial products and services, as well as implement extensive and reliable financial marketing research and efficient campaigns, to achieve and enhance customer loyalty and profitability.

The above goals can be achieved through the seamlessly integration of information technology solutions and business objectives at every process of the bank business that affects the customer.

3 The CRM Process Cycle in Banks

To realize the full value of their CRM initiatives, banks must explore and alter the processes by which customer information captured, analyzed, distributed and used to drive marketing strategy and action programs. An integrated CRM solution enables banks to set up a closed CRM cycle, composed of a set of continuous iterative processes. It manages the total end-to-end customer related process for a bank, collecting customer data, analyzing customer profile and life time value, supporting marketing decision making and optimizing the implementation of marketing campaigns, sales and customer service strategies across multiple channels throughout the bank (Pound 2000). The model of the CRM process cycle (Figure 1), based on a generic business view, presents a continuous improvement of the value between banks and their customers across touch points. The necessary stages are the following:

Customer data collection

Customer data analysis

Marketing strategy and action programs (planning, execution and effectiveness evaluation – customer response metrics)

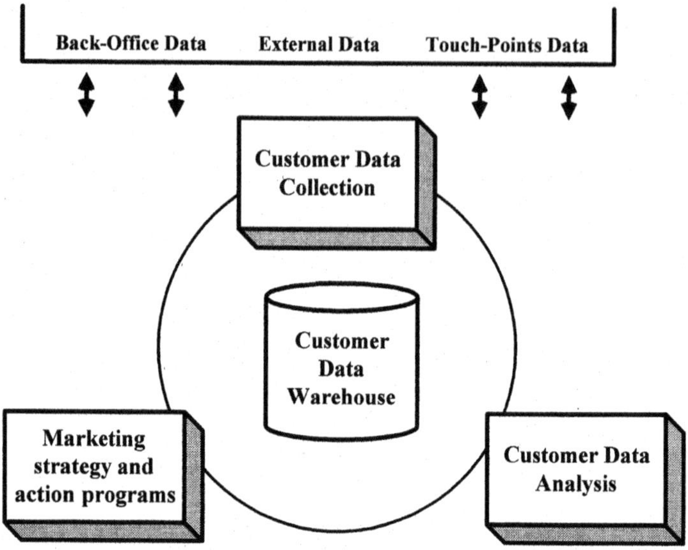

Figure 1: CRM Process Cycle in Banks

Today's banking data sources are highly heterogeneous. Information is geographically dispersed due to continual mergers, acquisitions and reorganizations. For example, a bank might use web site, email, ATMs, call centers, sales and marketing automation applications as stand alone systems, that must be integrated in a unified CRM banking environment. An effective, intelligent multi-channel customer interface will not be possible without a centrally-integrated warehouse driving the whole CRM process cycle. This must be updated in real-time and be enriched with back-office and external data to aid customer knowledge. It should also contain a record of historical data, which can be used to construct propensity models and customer lifetime value models to understand past behavior and actions in order to drive future marketing communication strategy.

A framework of an integrated CRM solution is presented in figure 2, with its main components: the front-end CRM system, the CRM support system, and the central customer data repository.

3.1 Data collection

Data in banking transaction systems (e.g. checking, savings, credit) are often organized around "accounts", "products", "channels" or other similar transactional concepts, that limit the ability to identify unique customers and their total relationship to the bank. One of the major goals of an integrated CRM system, is to consolidate these "information islands" and separate solutions, to form an open cross-bank system from which all executives, branch employee and business area departments officers, can share the same customer information. The necessary basis of data supply to an integrated banking CRM system can be obtained from:

operational (contact) sources, that is the customer interaction touch-points (branch, personal contact, ATM, call-center, internet banking, mobile banking, etc) (Chou, Chou 2000),

internal sources, that is the available information islands, from other bank's product-oriented systems and databases (deposits, cards, loans, investments, etc), meta-data analysis, reliable data mining results and marketing campaign response,

and from external sources, like marketing researches, infomediaries etc. providing geodemographic, lifestyle and psychographic data, which can help in developing the full customer image.

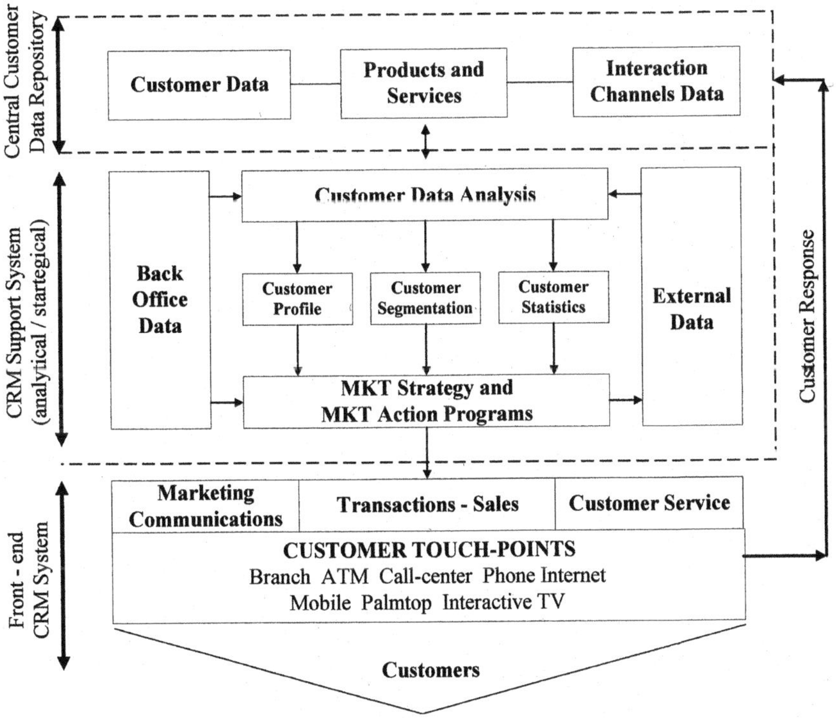

Figure 2: Framework of an Integrated CRM

3.2 Analysis

Deriving valuable customer insights from the data collected in a central customer warehouse requires simple or more sophisticated analysis techniques

(Heygate1998). Advanced analytics, including OLAP (Online Analytical Processing) tools and data mining techniques, extract relevant patterns or trends in the data.

Key integrated customer management insights provided by customer data analysis are customer differentiation / segmentation, distribution and concentration of customer value, share of purchase/ profit, and analysis of strategies that widen / lengthen / deepen customer relationship (Lawer 2000).

Customer data analysis enables the identification of customers profile and customer preferences for specific bank products and services, indicates the most appropriate channels to reach the customers, and assesses the profitability and life time value of every individual (Hawkes 2000). Furthermore, the future behavior of the customers can be predicted by analyzing their historical behavior. Customer statistics, customer profile and segmentation are the main outputs of the analysis stage feeding the marketing strategy planning and execution process. Having easily available information to marketing, makes the difference between a successful campaign and a failure (Delto 1998).

3.3 Marketing strategy and action programs

Captured data and results of customer analysis support marketers to refine marketing processes, messages and strategies. Marketing decision makers are provided with the tools and processes to discover the true value of data, converting information into usable customer knowledge deployed to enhance marketing decision making and CRM initiatives and campaigns.

Planning, execution and monitoring marketing strategies and action programs that personalize and optimize each customer contact is the goal of marketing automation within CRM.

It is critical for bank CRM not only to mine their data resources to uncover patterns and insights but also to operationalise this throughout the bank activities to turn the customer knowledge into value creating action.

The results from marketing and CRM activities and strategies continue the process knowledge acquisition enhancing the on-going value of marketing data intelligence, closing the feed back loop. Therefore, the final part of the CRM process cycle is the evaluation of the results of campaigns driven by marketing data intelligence. It is critical to measure performance and feed results back into the central customer data warehouse, in order to deliver greater customer intelligence in the future.

3.4 Data integration and maintenance

In the customer-centric data warehouse, the customer data from each transaction system will be integrated, to provide a unified view of the customer. Customer-centric data warehouses require complete, accurate views of customers, with their associated data, to feed customer data analysis and support marketing decision-making procedures (Joshi and Curtis 1999).

A CRM approach requires a new customer-centric business model, which must be supported by a set of applications coordinating the front office with the back office systems and e-commerce strategies (Bauer et al 1998, Fluss et al 2000). The integration of customer centric applications with the internal back-end systems and the creation of a central database of customer knowledge are believed to have a direct link to a more profitable bank organization. The customer makes contact with the enterprise either in a pre-sale, sale, or post sale situation as part of an ongoing relationship. Matching incoming with existing data, such as change of address, regular credit lines cancelled, complaints made, account balance decline, requires data warehouse transformation and cleansing tools in order to maintain an updated customer view (Eckerson, 1997, Carleton, 2000).

The use of a centralized relational database and open systems to manage customer data, contact history and relationship results allows the integration, at the data level, of the various technologies deployed at different stages in the CRM process. Data analysis models may be stored alongside the actual data, and value or profitability scoring and segmentation keys can be made directly available to marketing campaign management and marketing actions programs software. An integrated CRM system forms an open cross-bank system from which all executives, branch employee and business area departments officers, share the same customer information and are benefited from the same customer-centric intelligence. There is a holistic rather than partial view of the customer in the entire organization, and banks' executives are able to add value to the customer relationship, planning personalized marketing campaigns and appropriate product/service offerings.

4 Application Modules and Technologies

The key application modules, related to its main processes (customer relationship data collection, customer data analysis, marketing strategy and actions programs) and specific activities, as well as the technologies enabling the effective capture and analysis of massive amounts of data are presented in table 1.

Table 1: The banking CRM cycle's components

PROCESSES	ACTIVITIES	APPLICATION MODULES	ENABLING TECHNOLOGIES
Customer Relationship Data Collection	▪ Existing customers banking relationship data collection ▪ Potential customers general data collection	▪ Operational banking data capturing ▪ Legacy data capturing ▪ Historical customer behavior data capturing ▪ External data capturing	▪ Data base Warehouse ▪ On Line Transactions Processing (OLTP) ▪ Cookies ▪ Intelligent agents ▪ Web infrastructure ▪ Data repository ▪ Data cleansing ▪ Data refreshing
Customer Data Analysis	▪ Customer profile ▪ Customer statistics ▪ Customer life time value ▪ Customer segmentation ▪ Forecasting ▪ Channel preferences	▪ Data extraction ▪ Scoring ▪ Predictive modeling ▪ Profiling (life time value, preferences profitability, behavior analytics)	▪ On Line Analytical Processing (OLAP) ▪ Advanced algorithms ▪ Collaborating filtering ▪ Multi-dimensional databases ▪ Analytics ▪ Data mining ▪ Data marts ▪ Parallel Processing
Marketing strategy and action programs (planning, execution, evaluation)	▪ Customers loyalty strategy planning ▪ Banking products/services personalization ▪ Personalized banking campaigns planning ▪ Banking interaction planning ▪ Channel planning & evaluation ▪ Sales strategy planning (cross selling-up selling) ▪ Sales reports ▪ Customers claims/complaints reports ▪ Campaign effectiveness	▪ Marketing campaigns-transactions, sales, customer service and support automation ▪ Monitoring ▪ Reporting ▪ Measuring Customer Metrics ▪ Modeling ▪ Evaluation ▪ Redesign	▪ Campaign management software ▪ Optimization and automation software ▪ Data mining ▪ Data marts ▪ Neural networks ▪ Personalization servers ▪ Push technologies ▪ Time triggering

Customer relationship data collection application modules enable the fast, efficient and flexible capture of massive amounts of existing and potential customer data derived from multiple sources, such as data from bank-customers' interactions, legacy, historical and external data (Chen and Frolick 2000). The enabling technologies concern the appropriate data cleansing and refreshing, to provide consistency despite multiple sources and to reduce data fracture and duplication. (SAS, 2000).

Customer data analysis application modules facilitate the analysis of large amounts of customer relationship-related data to produce information content relationships. Data warehouse-related technologies, the heart of any customer-centric banking system, enable the integration of all databases that are required to perform a comprehensive customer information platform. Data mining-related technologies automate the searching of large volumes of data, looking for patterns that accurately predict customers' profiles. On-line analytical processing (OLAP), allows bank's executives to derive information and customer intelligence from data warehouse systems by providing tools for querying and analyzing data leading to multidimensional view of the customers.

Marketing strategy and action programs application modules, provide the appropriate tools, to support marketing campaign planning, running, monitoring, reporting and evaluation. Designing marketing action programs that personalize and optimize its customer contact is the goal of marketing campaign, sales force, and customer service and support automation modules (Cherry Tree 2000).

Banking marketing campaigns automation, facilitate executives to determine customers and market profiles, identify customers loyalty and profitability, plan and implement specific marketing campaigns, select appropriate contact channels and track customer responses across these channels (Robicheaux and Coleman 1994).

Banking sales force automation, comprises technologies that automate sales-related banking activities especially sales forecasting, planning, contact and quote management and sales analysis. Decision tools supporting activities of cross-selling, up-selling, selling functionality enhancement and selling convenience achievement, are also included in sales banking sales force automation.

Banking customer service and support handles all types of customer queries including banking product concerns, information needs, requests and banking services.

5 CRM Architecture

A complete customer centric view requires the capture and storage of customer transactions and marketing campaign responses across disparate touch points, and additionally customer data from back-end transactional systems and external sources. A centralized customer data warehouse with a reliable, scalable and highly available storage infrastructure gives the solution to the problem of data consolidation and integration of diverse customer data assets. A customer- centric data warehouse requires complete, accurate views of customers, with their associated data in order to support the bank's decision-making process.

Figure 3 depicts the basic components and architecture of an integrated banking CRM solution.

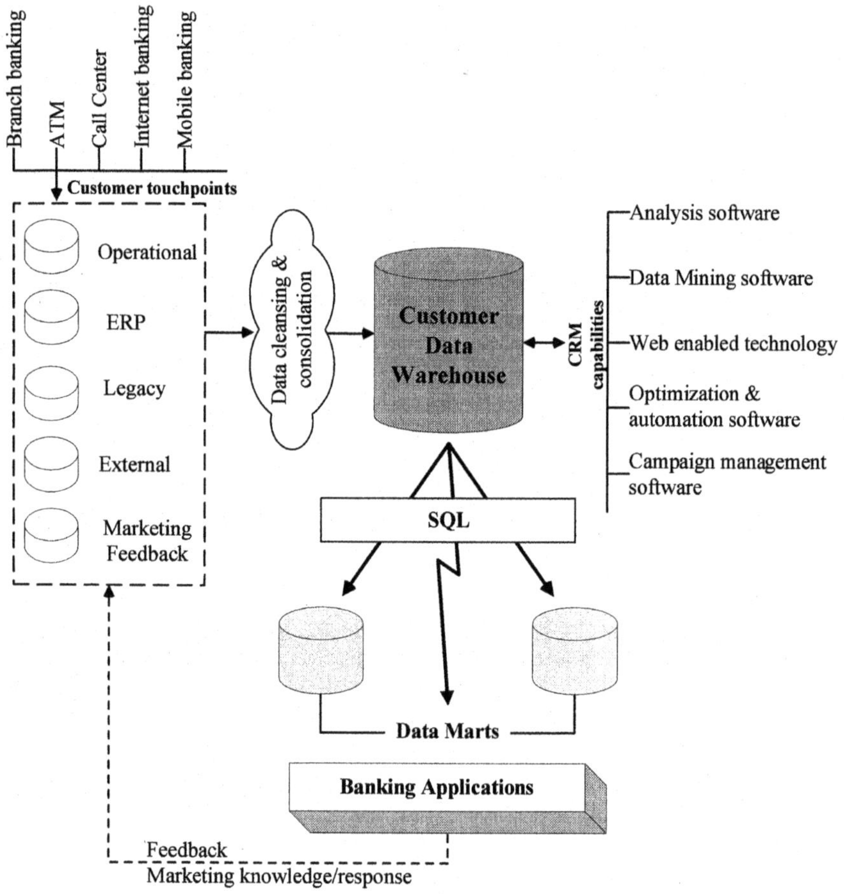

Figure 3: CRM Architecture

An integrated banking CRM invests on the knowledge discovered through the use of CRM capabilities – data warehousing, data mining, customer analytics, optimization and automation software, online analytical processing (OLAP), campaign management software and web enabled technologies – to develop a customer oriented marketing strategy and applications (Martin 1998). Data mining is the process of selecting, exploring, and modeling large quantities of data in order to uncover previously unknown meaningful patterns and rules (SAS, Firestone). The goal of data mining use in a banking CRM solution is to allow the bank to improve its marketing, sales, and customer support operations through better understanding of its customers (Below 1998). On-line analytical processing, allows bank's executives to derive information and customer intelligence from data warehouse systems by providing tools for querying and analyzing data, leading to multidimensional view of the customers. After mining the customer data, bank executives feed the results into campaign management software, which in turn automates the planning, execution, and evaluation of targeted marketing action programs adjusted to customer expectations. Analysis software should be delivered as an extensible application that uses a set of customer profile and profitability models and reports. Customer analysis results should integrate with campaign management and automation software to enable banks to transform customer information into building better customer relationships. More informed banking decisions can be carried out with the aid of analytic tools that consolidate and manage the vast amount of customer knowledge within the bank. Web enabled technologies support bank organizations to treat the e-channel and e-business as part of overall CRM strategy to avoid channel conflict and optimize the return on customer relationship.

A data mart, in the context of a CRM system, is a decision support system incorporating a subset of the customer's data focused on specific banking applications or activities. Data marts allow for greater flexibility or increased performance. However, the data mart must be incorporated into the overall customer data warehouse architecture and managed and populated from this central data warehouse (table 2).

Table 2: Data marts organization

Specific Customer Profile	**Specific Product/ Service**	**Specific Channel**	**Specific Marketing Action Management**
▪ Existing ▪ Prospects ▪ Individual ▪ Corporate	▪ Consumer banking ▪ Retail banking ▪ Business banking ▪ Corporate banking ▪ Investment banking ▪ Bancassurance	▪ Branch ▪ ATM ▪ Call center ▪ Phone ▪ Internet ▪ Mobile ▪ Interactive TV ▪ Palmtop	▪ Marketing campaigns ▪ Sales ▪ Customer service & support ▪ (planning, execution, reporting, evaluation)

Specifically, in relation to banking CRM the main areas for data marts organization are the specific customer profile, the specific product/service, the specific channel, and the specific marketing actions-management (Magic Enterprises 2000).

6 Conclusions

An integrated CRM is a banking strategy aimed at understanding and anticipating the needs of a bank's present and potential customers. From a technological perspective it involves the customer data capture from across the enterprise, and the consolidation of customer related data in a central database. Allowing knowledge workers to access and analyze the consolidated customer data is the additional demand on an integrated CRM system.

The primary goals of an integrated banking CRM system is to increase customer retention, personalize interactions with customers at every point of contact, reduce costs and increase value of customer interactions, thereby maximizing profits.

According to the model demonstrated in this paper, a new "customer-centric" banking strategy is followed. Technological tools and CRM capabilities improve the operation, effectiveness and performance of the banking sector. Customer knowledge is derived from a central customer data warehouse and specific data marts, allowing the bank executives to implement customer tailored solutions in order to increase customers' loyalty and profitability.

An effective integrated CRM banking system integrates key CRM processes and related activities to the operational and strategic core banking systems and applications. Particularly, it addresses customers in a common way in the entire bank, reevaluates both CRM and corporate banking processes and functions, and implement and maintain an integrated data warehousing system with common and validated information on customer profiles, campaign activities and sales contacts, regardless of the procurement channel.

While information technology plays a vital role in a banking CRM solution, it should be driven by a strategic management perspective.

The promise of CRM is significant but banks must realign their business processes and culture to a more customer-driven model through the integration of disparate people, processes and technology. A deeper examination concerning the problems appearing during the implementation stage of CRM in the banking sector is necessary and calls for more research within this field.

References

Bauer, H., Grether, M., Leach, M., (1998), "Kundenbeziehungen ueber das Internet", Dermarkt, jg 37, pp 119-128

Below, von C., (1998), "Data Minining", in: Managementinformations-systeme in Marketing und Vertrieb, Hrsg. Uwe Hannig, Stuttgart, pp. 53-62

Carleton Corporation, 2000, "The Four Challenges of Customer-Centric Data Warehousing", www.crm-forum.com

Chablo, E., 1999, "The Importance of Marketing Data Intelligence in Delivering Successful CRM", www.crm-forum.com/crm_forum_white_papers/mdi

Chen, L., Frolick, N., (2000), "Web-Based Data Warehousing, Fundamentals, Challenges and Solutions", *Information Systems Management*, is. Spring 2000, pp. 80-86.

Cherry Tree and Co, (2000), "Extended Enterprise Applications", Spotlight Report, www.Cherrytreeco.com

Chou, D.C., Chou, A.Y., 2000, "A Guide to the Internet Revolution in Banking", Information Systems Management, spring, pp. 51-57

Delto, A. (1998), "Kundeninformationen professionell nutzen mit Customer Relationship Management", in: Managementinformations-systeme in Marketing und Vertrieb, Hrsg. Uwe Hannig, Stuttgart, pp. 83-92

Eckerson, W., 1997, "Finding the Dividing Line Between Data Cleansing and Data Transformation Tools", 2:3, pp 30-32

Fluss, D., Amuso, C., Hope-Ross, D., (2000), "Internet-based customer service: Miracle or migraine?", *The Journal of Customer Loyalty*, is. 14, pp.14-23

Hawkes, V.A., 2000, "The Heart of the Matter: The Challenge of Customer Lifetime Value", www.crm-forum.com

Joshi, K., Curtis, M., (1999), "Issues in Building a Successful Data Warehouse", "*Information Strategy: The Executive's Journal*, is. Winter 1999, pp. 28-35.

Lawer, C., 2000, "Integrated Customer Management", www.crm-forum.com

Magic Enterprises, 2000, "The CRM Phenomenon", www.magic-sw.com

Martin, W., (1998), "Effektives Marketing durch Data Warehousing und Data Mining", in: Managementinformations-systeme in Marketing und Vertrieb, Hrsg. Uwe Hannig, Stuttgart, pp. 33-46

Miller, H., (2000), "Managing Customers Expectations", *Information Systems Management*, is. Spring 2000, pp. 92-95.

Peter, S. (1998), "Beziehungsmanagement: Kundenbindung als Marketingziel", Absatzwirtschaft, 40:7, pp 74-82

Pound, J., 2000, "Making CRM Work", www.crm-forum.com

Robicheaux, R, Coleman, J., (1994), "The Structure of Marketing Channel Relationships", Journal of the Academy of Marketing Science, 22:1, pp. 38-51

SAS Institute Inc. (2000), "Taming the CRM Information Explosion", www.sas.com

Sheth, J.N., Parvatiyar, A, (1995), "The Evolution of Relationship Marketing, International Business Review, 4:4, pp 397-414

Vlachopoulou, M., (1999), e-Marketing: Information Systems – New Technologies in marketing , Rosili, Athens

Bank Mergers, Acquisition and Concentration

Mergers and Acquisitions in the European Banking Sector: The Case of BNP and Paribas

J. Gupta, A. Chevalier

ESCP-EAP
79, avenue de la République
75543 Paris Cedex 11
FRANCE

Summary:
The financial sector is undergoing rapid restructuring through mergers and acquisitions operations at the global level. The European banks are following the global trend. The paper analyses the takeover battle between three French banks; the BNP, Société Générale and Paribas. The purpose of the analysis is to understand the logic behind this operation within the context of the financial theory. The paper includes the profiles of the three banks involved in the operations and the chronology of the events leading to the final outcome and critically analyses the proposed synergies. The importance of the method of payment and its impact on the operation is also discussed. The conclusions are drawn on the real motivations behind these operations.

Key Words: Merger and Acquisition, Banks, Synergy, Takeover

1 Introduction

In the last few years (1993-2000), the banking sector has witnessed tremendous Mergers and Acquisitions activity. This has led to a spate of consolidation in the sector and the European banks have not been to far behind in their moves for consolidation. This has led to the creation of giants such as Deutsche Bank, UBS, HSBC and Crédit Agricole in Europe. The moves for consolidation are continuing and it extends to Insurance sector; the latest example being the case of the German bank Dresdner and the German Insurance company Allianz (2001). The table 1 below gives the list of the world's largest banks in terms of the total assets in 1998 (before the BNP/ Paribas merger). We notice that three of the worlds' top banks (in terms of total assets) were European.

Table 1: Worlds' Largest Banks

Rank	Bank	Country	Total Assets($bn)
1	DKB/Fuji/IBJ	Japan	1,250
2	Deutsche Bank	Germany	735
3	UBS	Switzerland	684.5
4	Citigroup	USA	668.6
5	Bank of America	USA	617.7
6	HSBC Holding	UK	475.6
7	Credit Agricole	France	455.8
8	ICBC	China	391.2
9	Chase	USA	365.9
10	Bank One	USA	256.0

According to the financial theory, when a firm buys another company, it is a capital investment and the basic principles of capital decisions apply. The acquisition decision should contribute to the shareholders' wealth. The difference, between a classic capital budgeting and an acquisition decision is that in the second case the shareholders of two separate firms are involved and it is logical that the wealth of the shareholders of both the firms is increased. In other words the synergy gains from the merger operation should be shared between the shareholders of the two firms (Myers).

The economic gain of a merger is obtained by adding the present values of the excess cash flows of the merged firm as compared to the cash flows of the two firms if they operated as independent firms. The discount rate should be the cost of capital of the merged firm. Therefore, a merger operation is only justified if the two firms together are worth more than apart. The economic gain is given by:

$$\sum \frac{CF_t(AB)}{(1+r_{AB})^t} - \sum \frac{CF_t(A)}{(1+r_A)^t} - \sum \frac{CF_t(B)}{(1+r_B)^t}$$

Where

$CF_t(AB) =$ Cash flow for the period "t" of the merged firm

$r_{AB} =$ Expected Cost of Capital of the firm after merger

$CF_t(A) =$ Cash flow for the period "t" of the firm A apart

$r_A =$ Expected Cost of Capital of the firm A apart

$CF_t(B) =$ Cash Flow of the period "t" of the firm B apart

$r_B =$ Expected Cost of Capital of the firm B apart

The economic gain is then shared between the shareholders of the two firms. The synergy gains can be due to increased cash flows or decrease in the cost of capital or eventually a combination of both. We notice, however that estimating the energy gains is extremely difficult, particularly so, in the case of banks. Studies (Jensen and Ruback) have been carried out to see if there are any positive abnormal returns to the acquiring firms' shareholders and to the selling shareholders. The results are not conclusive, in general findings show that the returns tend to be positive for the selling shareholders. The financing method used (cash or stock), has an influence on the returns. This is explained by the asymmetric information between the management of the acquiring firm and the acquired firm(Asquith, Bruner and Mullins). The importance of the form of payment has been discussed by Jensen and Murphy. A comparative study of M&A operations in the European context was carried out in 1990(Chevalier and Gupta).

We feel that each M&A operation is to a certain extent unique, because the econometric studies over a period of time do not capture properly the specific factors linked to the economic environment and more specifically the importance of the behavorial and domino effects. We feel that an in-depth case study approach is useful to capture the different dimensions of a M&A operation (Ruback 1983).

The objective of this paper is to provide an understanding of the logic behind the takeover battle involving BNP (Banque National de Paris), Société Générale and Paribas, three French Banks during the year 1999.

The nature of the operations involved were complex and it has been our endeavor to lay out the facts as clearly and precisely as possible. The information and data were obtained from published sources and also through interviews of the actors involved in the operation. It should, however be stressed that most of the information in M&A operations are confidential and the parties are often reluctant to provide them.

The rest of the paper is divided into six sections, the second section analyses the reasons for the recent activity of mergers in the banking sector, the third part discusses the reasons for mergers in the European banking sector, the fourth section gives the profiles of the three banks involved and the chronology of the events, the fifth section carries out a critical analysis, the sixth section comments on the proposed synergies and finally we draw some conclusions.

2 Reasons for the Recent Activity of Mergers in the Banking Sector

2.1 Economies of Scale and Scope

Size does matter. By consolidating their activities with others , banks hope to gain economies of scale. These economies of scale become particularly important for investments in information technology systems. Installation of IT systems is becoming extremely important in this new era of technology. The IT systems will enable banks to offer better products to their customers in the face of increased globalisation in today's financial markets. Increased size and the associated economies of scale enable banks to offer more products.

2.2 Fragmentation

The banking industry is very fragmented across the world. For example, in Germany, the four largest banks hold less than 20% of the retail market. This fragmentation coupled with over capacity in the banking sector has created a climate that is ideal for mergers and alliances.

2.3 Increased Competition

Increased competition in the banking sector has put a pressure on the margins. This new competition forces banks to price loans with such small margins that the risk is not adequately covered. Thus although banks wish to increase their size, they do not want to add capacity to the industry. Thus, the best option is to have a merger or an alliance.

2.4 Diversification

Some banks are using alliances and mergers as a strategic method to establish a platforn in previously untapped markets. An excellent example is the take over of Bankers Trust by Deutsche Bank. Through this move, Deutsche Bank hopes to make inroads into the American Investment Banking sector. between developed and developing countries.

2.5 Cost Cutting

Often there are many overlaps in the operations of two banks especially in terms of the geographic area covered. By merging, the size of the bank increases, but tremendous cost savings are to be had by closing redundant retail branches, dealing rooms, expensive IT systems, and of course by laying off employees.

2.6 Creation of Shareholder Value

A traditional reason for a merger in any industry is the failure of one company's management to deliver value. The same holds true for the banking sector. Bank of Scotland launched a hostile $36 billion bid to take over National Westminster Bank. Bank of Scotland promises to deliver savings worth $1.67 billion in a period of three years by cutting costs at National Westminster Bank which is considered the country's least efficient bank.

2.7 The Domino Effect

Finally, another reason that banks are merging is simply because everyone else is doing it! The formation of Citigroup, the continuous expansion of Deutsche Bank, and the collapse of the Japanese financial sector, triggered the merger involving Dai-Ichi Kangyo Bank, Fuji Bank and Industrial Bank of Japan.

3 Reasons for Mergers in the European Banking Sector

The value of European bank-based M&A activity soared by a third to $162 billion in 1998[1]. This is in line with global trends in the banking sector. Apart from the reasons explained earlier , there are a few points specific about the European banking sector. Increasing international competition, the convergence of the European market and growing pressure on costs and margins are forcing European banks to fundamentally rethink their strategy and to reconsider their position in the market.

Europe already accounts for three of the world's ten largest banks ranked by assets[2], but the creation of powerhouses such as Citigroup, and the planned three-way tie among Fuji Bank, DKB and IBJ has placed pressure on the Europeans to merge to create bigger banks.

The deregulation of financial services in the European Union (EU) and the launch of the Euro have unleashed a wave of merger activity in every EU nation. A common currency will boost competition across Europe since products will be priced in the same currency. Companies with high costs will get squeezed.

---

[1] Source : Thomson Financial Securities Data

[2] Refer Table 1

Banks have faced very little opposition from regulators because of the fragmented nature of the banking sector. The European market still shows a low concentration level (10% for all bank categories and 15% for Universal banks as compared to 18% for US banks). Even the merger of the first and third-largest Spanish banks earlier this year passed regulatory measures. Ordinarily, this would have met with tremendous resistance due to monopolistic concerns.

As Europeans are saving more for retirement, bankers and insurers are joining forces to peddle products from life insurance to mutual funds.

One interesting trend is the focus on domestic rather than cross-border deals, with the latter accounting for just 11 percent of this year's deal total, according to Thomson Financial, this phenomenon maybe attributed to the fact that domestic deals offer more scope for cost savings and circumvent the potential for cultural conflicts. But this trend is not going to last for too long. The creation of the European Monetary Union (EMU) is forcing Europe to rethink the notion of 'local' and 'national' banks. As this region becomes more and more homogeneous , companies will face fewer advantages on the domestic market over foreign competitors and their current market position wil be more under attack. This in turn will increase the importance of being present on the European market. The removal of long held priveleges on the domestic market will promote cross border mergers and relieve the national consolidation processess that still dominate the European banking world.

4 Profiles of BNP, Societé Générale and Paribas[3] (Pre-merger Situation)

4.1 BNP

Presently, BNP is France's most profitable bank. Since its privatisation in 1993, BNP has been successfully carrying out a modernisation program for the benefit of its stockholders, customers and employees. Since 1993, the net income attributable to the bank has increased sevenfold to about FF 7.3 billion and earnings per share by a factor of 5, to FF 33.86. Its return on equity has increased from 2.2% at the end of 1993 to 11.8% at the end of the year 1998. This has made BNP France's most profitable non mutual bank.

---

[3] Information regarding the history and financial performance of banks has been taken from their internet sites

Nature of activities

Retail banking

Within France

This core business comprises the retail banking activities of BNP and Banque de Bretagne, as well as logistics, marketing, collection, and other support functions provided to those entities. Serves 6 million individuals.

Outside France

The core business outside France, comprises BancWest (the fifth-largest bank in California and the second largest in Hawaii, with 220 branches and 800,000 customers) and BNPI-SFOM (330 branches and 500,000 customers in Africa, the Middle East, and the Indian Ocean). BNPI, another subsidiary also handles the BNP Group's business in France's overseas areas.

Private banking and Asset management

The private banking wing of BNP is involved in designing portfolios according to the needs of both individuals and companies. It uses the information and resources of their economic research wing to analyse the investments of its clients and to recommend alternate and better investments. Asset management at BNP involves centralising the management of assets, achieving optimum tax efficiency and shielding assets from possible exchange controls.

Investment banking

Banexi, a fully owned subsidiary of BNP carries out the investment banking activities of the group. These activities include : Mergers and acquisitions, Financial engineering, Privatisation of companies and Valuation.Banexi is present in fifteen countries outside France. BNP Banexi is the world's largest merchant bank in number of transactions and is the fourth largest bank in terms of value of transactions.

Electronic banking

This activity, known as BNP Geolink teletransfer is presently focussed on Centralising of accounts worldwide, Centralised funds transfer and e-banking.

Export finances / Trade services OR wholesale banking with structured Finance

The structured finance division of BNP is involved with the design of financial instruments to reduce risk. These risks arise from two sources, namely interest rate variation and exchange rate variation. They are involved with the design of SWAPs, FRAs, collars and other options.

Specialised financial services

BNP Lease a fully owned subsidiary of BNP has been organised as the BNP World-wide line of Business for leasing in December 1998. The International Division of BNP Lease has been set up to that effect, and it is fully operational since the beginning of 1999. It has 20 branches in 15 countries. It provides lease services in capital goods and real estate except those required by the aeronautics and shipping industry that is done by other group companies.

Economic research

The economic research wing of BNP is like the R&D division of a manufacturing organisation. The results of this unit are used for their activities in structured finance, export finance and merchant banking.

Income growth of BNP

As mentioned earlier, the income of BNP has grown sevenfold from 1993. The details of income, EPS and return on equity growth of BNP in millions of Euro are given in Table 4.

4.2 Société Générale (SG)

Société Générale is a prominent French bank. Société Générale is the largest private sector bank in France, with assets of about $412 billion, BNP is number two at about $349 billion. While Pariabs has assets worth $284 billion. The figures correspond to year 1999. Société Générale had the sixth largest market capitalisation on the Paris Stock exchange amounting to 22.25 bn Euros at the end of November 1999. It has a good presence abroad with as many as 500 branches outside Fance though most of its activities are concentrated in France. (2,600 branches).

Nature of Activities

Société Générale's business is broadly divided into Retail Banking, Asset Management and Private Banking, Corporate and Investment Banking, and Corporate Centre. The division of Income across these areas for the first half of 1999 is indicated in the *Table* 3:

Table 3: Division of income across varions activities

Area	*Income (EUR Million)*	*Percentage of Total (%)*
Retail Banking	93	30.7
Asset Management and Private Banking	88	6.9
Corporate and Investment Banking	3 467	36.4
Corporate Center	333	26,0
Total:	1 281	100.0

Société Générale offers a wide range of financial services both to individuals and to institutions:

For small and medium sized businesses, the bank offers Equipment Finance, International Trade Finance, Electronic Banking Services, Employee Benefit Plans, Capital Markets Services, Working Capital Management and Finance, Corporate Finance, and Investments and Deposits.

SG aims to act as a focal point for multinational firms and provides services in commercial, transaction and investment banking. More specifically, it provides advisory services for Mergers and Acquisitions, consulting for primary and secondary equity linked transactions, and the whole gamut of services related to raising finance for businesses.

SG also act as an interface to the financial markets, including equity based derivatives, fixed income securities, commodities, and currencies. It is also active on the futures markets.

SG Asset Management has more than 35 years of experience in securities with more than FF 865 billion in assets under management.

SG also provides international cash and liquidity management services.

For institutional investors, SG has a department, comprising of specialists in key fields such as insurance, pensions and fund management.

For high net worth individuals, SG has wealth management and private banking service.

Société générale is actively involved in developing financial instruments for hedging and arbitraging. It is also active in the for ex and interest rates markets. Its capital market activities employ as much as 4500 people representing almost 15 % of its total workforce. The key figures of the bank Société Générale (SG) are given in table 4.

4.3 Paribas

Paribas was founded in 1872. It had started of, as a bank with an industrial focus. It had been a major player during the two world wars. It started penetrating the world market in the 60s. It was subsequently nationalised and then privatised in the 80s. Paribas then took up a restructuring exercise in the 90s with a focus on core areas. It has 220 branches in 60 countries worldwide.

Paribas has a shareholders' equity (1998) of FRF 56.4 billion (USD 8.6 billion) and a Cooke solvency ratio of 9.1%. Paribas's aim is to strengthen return on equity - currently at 12% - to 15% by the year 2000.

Nature of activities

Paribas is a specialised international bank with a strong European base. Its strategy is focussed on selective growth in expanding markets with emphasis on technological innovation, creativity and world-wide presence (presently active in 60 countries) and distribution.

Paribas is organised around four core businesses:

Corporate and Investment Banking

Asset Management and Banking Services

This includes the Institutional and Private Asset Management activities of the various subsidiaries of Paribas such as Banque Paribas, Cardif, Cortal, Global Securities Services, and Banque Directe.These have recently been combined within single asset management and services sector.

Its clients continually recognise Paribas's position as the leading provider of securities services in Europe. In the annual Agent Bank Survey for Major Markets published by Global Custodian magazine, Paribas received a "Top Rated" or "Commende » status in all markets where it operated.

Banque Direct, a home Banking system , is an innovation by Paribas which enables clients to conduct their banking business at the most convenient time, in absolute safety, and under competitive terms.

Retail Financial Services

Within Paribas, dedicated subsidiaries like UFB Locabail, Arval, Cetelem, and UCB provide specialised financial services to individuals and businesses.

UFB Locabail provides capital good financing and factoring services to a clientele of European businesses. Arval, a subsidiary of UFB Locabail, provides long-tenn rental of vehicle fleets. The European consumer finance leader, Cetelem and its subsidiaries offer durable goods, automobile and home-improvement loans through 102 branches, direct marketing, and the outlets of its retailing and financial partners, including Carrefour, Conforama, Darty, the French savings banks and Banques Populaires networks.

Proprietary Investments and Real Estate

This includes: Paribas Principal Investments (PPI), one of the largest and most active equity in Europe (with a portfolio of EUR 6.5 billion), Klépierre (one of the largest listed commercial real estate investment companies in France), Ségécé (major player in the development and management of shopping centers in France) and Sinvim (a real estate development company).

The comparison of key statistics of the three banks is given in table 4.

Table 4: Comparison of key statistics

	BNP	SG	PB	BNP	SG	PB	BNP	SG	PB
Total Assets (bil Euro)	283.7	271.1	232.1	310.2	374.9	210.2	324.8	383.5	248.5
Net Income (M Euro)	588	693	663	909	993	1002	1114	1073	1002
EPS (Euro)	2.85	7.94	-	4.31	9.67	3.90	5.16	10.73	3.92
ROE %	7.4	8.7	-	10.40	10.6	13.0	11.8	11.2	12.0
Pruden-tial Ratio %	9.1	10.22	-	9.9	11.11	8.9	10.4	11.15	11.0

Source: Annual reports

4.4 Chronology of events[4]

Presented below is a sequence of events that led to the culmination of the take-over battle between BNP, SG and Paribas.

1998-1999: SG & Paribas had plans of a merger. These plans however, were not shared with the public.

Feb 1, 1999: SG announced that it will buy Paribas for a share exchange for 15.1 billion Euro. Such a merger would lead to the creation of France's largest bank.The details involved were as follows:

Exchange ratio of 5:8 for every Paribas Share

17% premium for Paribas

Feb 15 , 1999: An attack by major European banks especially BNP was predicted by the market.

[4] Source : M&A Daily

March 9, 1999: BNP announced its plans of creating SBP by taking over SG & Paribas for $37 billion Euro's. the market response to this offer was favorable. This would lead to the creation of the worlds largest bank with assets in excess of 1 trillion USD. The other details were as follows:

11:8 for Paribas (18.2% premium) and 15:7 for n (14% premium)

SG & Paribas termed the offer as 'unfriendly' and 'hostile'

BNP's Chairman, Michel Pebereau said that he was ready to merge with Paribas alone

There was a fear of thousands of job losses

March 1 Oth - June 14th , 1999: All along, a counter offer from other European banks was expected.

June 14, 1999: SG sweetened its bid for Paribas

5:8 + 75 Euros OR 3:2 (for first 30%)

June 21, 1999: The Commission for Lending Institutions (CECEI) denied authorisation for SG saying it required more deliberation.

June 23, 1999: A meeting of the CEO's of the three banks with Chairman of Bank of France, Mr Trichet

Mr Trichet proposed splitting Paribas 50:50 through a holding company

June 29, 1999: No solution was found. There was a total collapse of talks.

July 1, 1999: BNP improved its offer for both SG and Paribas. The increase figure amounted to $41 billion.

July 7, 1999: CECEI gave BNP authorisation to "take control of 50.01% or more. It also authorised a higher bid for SG too.

July 8, 1999: SG hinted at increasing its bid for Paribas for a second time.

July 12, 1999: SG rejected the improved offer of BNP

July 22, 1999: Axa, Europe's largest insurer said that it will tender 7.4% in Paribas and 0.4% in SG to BNP. AGF, French insurer with 5.09% and 2.4% rejected BNP offer.

Aug 13, 1999: BNP gets 65% share in Paribas but only 31.5% (significant minority) in SG. This lead to chaos; however, SG stock soared.

Aug 16, 1999: The regulator was to decide the three-way deal after SG shareholder's rejected the BNP offer. The merger between BNP and Paribas goes through.

Aug 19, 1999: Trichet proposed that SG and BNP hold comparable stakes in each other.

Aug 22, 1999:SG was not averse to working an alliance with BNP in strategic areas but refused any stake.

Aug 23, 1999: Pebereau said that SG will be allowed to continue as a separate entity for three years, with BNP maintaining its stake. SG rejected this offer.

Aug 24,1999: Chairmen prepare to meet with the CECEI ;

Aug 25, 1999: Paribas chief, Levy Lang quit after failure of merger with SG.

Sept 21, 1999: BNP decided to launch bid for the rest of Paribas. Chairman Michel Pebereau said that a merger with Dresdner bank of Gernany was possible in the future.

5 The Details of the Battle: A Critique

The first seeds of the of the take-over saga, as mentioned in the chronology, were sown in 1998. The consolidation in the European Banking industry had brought Société Générale and Paribas to the negotiating table in 1998. As far as Société Générale was concerned, there were several strategic reasons for this move. The two banks worked in différent areas. Paribas is primarily known for its investment banking activities. Société Générale identified three core areas to focus on. These were retail banking, investment banking and asset management. It wanted to increase its focus on retail banking where it felt it would be able to do a better job. However, they were not interested in letting go of the investment-banking arm either because of the tremendous growth potential they saw in this area. Thus, Paribas was an ideal choice. In the field of investment banking too, both banks had complimentary skills. While Paribas had traditionally been strong in bond trading, Société Générale was strong in project finance, equity and the like. Société Générale also had to please analysts who complained that there was too much exposure to emerging markets in investment banking instead of the more stable and also profitable retail banking.

Paribas on the other hand knew that some form of consolidation or the other was bound to occur in France. The three most important banks in France are BNP, Société Générale and Credit Lyonnais. There was a danger that some form of strategic alliance might have been worked out between these three, all though they had been staunch competitors. Any such move would have hit Paribas. Staying independent in the era of globalistaion would not be beneficial for Paribas because sooner or later it realised it would be gobbled by a giant bank, probably a foreign one. Paribas also would have been vulnerable to a hostile take over with its smaller size. Although Paribas was on a faster growth path than the other two banks there was a question regarding the extent to which this growth could be sustained.

Thus the stage was set for the beginning of the battle. On February 1, 1999 Société Générale Chairman was supposed to meet the BNP Chairman Pebereau for discussions where they were supposed to discuss a possible alliance. This meeting was annulled and in a meeting next day , the Société Générale made its offer for Paribas public.

The merger of Société Générale and Paribas would have created France's largest bank. Their combined assets would have been approximately 600 billion Euros as against 315 billion Euros of BNP. The activities of BNP and Société Générale run neck to neck in the retail-banking sector. A merger between the two would have created a one-stop shop for customers who want one organisation to provide all the services. Paribas was growing at a faster rate than both BNP and Société Générale. The sales growth, profits and profitability of Paribas were also higher . If BNP did not make a counter offer, Société Générale alone would have reaped all the benefits of a fast growing profitable bank.

Société Générale had already formed a team to carry out a valuation of Paribas. Teams of Société Générale officiels were touring the world to convince share holders of the benefits of the merger. However, on February 1, 1999 when Société Générale came out with its first offer, the stocks of Paribas, Société Générale and BNP fell. Although it is difficult to give a precise reason for this kind of valuation by the market, there are some possible leads.

Shareholders did not feel that the proposed synergies were realistic

Unlike Société Générale, Paribas was doing very well as an investment bank. Shareholders felt that this core competence would be undermined when Société Générale takes over control of Paribas. Société Générale shares fell because investors already felt that Société Générale should focus on retail banking. An acquisition in an investment bank through exchange of shares was perceived to be counter-productive.

It is possible that shareholders saw a counter bid coming from BNP, but did not know for sure whether it would succeed. In the event it failed, shareholders would be left with shares of the new Société Générale Paribas bank.

The Société Générale offer of 5:8 for every Paribas share entailed a premium of 17%. The total value of the transaction was $17.86 billion. They predicted post tax synergies of approximately 602 million Euros in the year 1998 itself. It is however not clear how they expected to integrate and gain synergies in such a short time. The market too found it hard to believe.

As expected on March 9, 1999 BNP launched its counter offer. It announced its plan of creating the entity SBP by taking over Société Générale and Paribas for $ 37 billion. It promised an 11:8 ratio for Paribas (18.2% premium) and 15:7 for Société Générale (14% premium). BNP stock soared after this announcement. They felt that this deal would get synergies of $1.38 billion by 2002 itself. There were several likely reasons for the optimistic market sentiments.

BNP and Société Générale have very similar portfolios. The greatest source of value in such a merger comes from selling of large parts of the retail bank channel. The two banks virtually have a parallel chain of banks.

With similar activities, the IT infrastructure of both banks could have been easily used to cover the operations of an integrated banking system.

For the same reason there would have been substantial scope for man power reduction

BNP would never be vulnerable to a cross border bids anymore. An arrangement like this would also bode well for the French Government which would then facilitate the smooth consummation of the merger.

The other minor sources of value creation included revenue enhancements by providing a one stop solution, bargaining power in the money markets and sharing of best practices.

Paribas and Société Générale both rejected the offer. Société Générale had several reasons.

The merger had been initiated by BNP and though it was made to look like a merger of equals it would not have been the case. BNP had launched the bid specifically to scuttle the Société Générale effet. This coupled with the unannounced declaration of the merger bid by Société Générale would have entailed strained relations between the top management of the two banks.

BNP also said that they would have minimal job losses in the merger. Given the nature of the businesses of the two banks, this seemed unlikely (It was predicted that there would be approximately 6000 job losses if BNP and Société Générale merged) and such a step would definitely not add value to the shareholders. It is also apparent that Société Générale with its lower profitability would have had to bear the brunt of the layoffs when they actually took place. The top and middle management at Société Générale wanted to keep their jobs.

The top management of Paribas too wanted to merge with Société Générale. The specific reasons for this are not clear. The authors attempted to contact the officials of Paribas, but in the light of their resistance in the past and the uncertainty regarding the future they were not willing to offer their comments. There is however reason to believe that the organisational structure that would have come up as a result of this merger suited Paribas top management.

On June 14, 1999 Société Générale sweetened its bid for Paribas. The new bid, 5:8 + 75 Euros or 3:2 (for first 30%) immediately raised the stock prices of Société Générale. The market response to this move was very favourable. After accounting for the Beta of Société Générale, it can be seen that there was an 11% additional return on the Société Générale share. There seemed to be several reasons for this:

It was felt that BNP would now come up with an even better offer. It therefore made sense to have a stake in Société Générale. When the first bid was made, the shareholders knew that BNP would bid but were not sure whether it would succeed or not. This time however, the market had started feeling that BNP was very determined. The resulting synergies would bode well for all shareholders.

Arbitrageurs began to buy shares of both Paribas and Société Générale on realising that a fierce boardroom battle lay ahead. This provided an upward thrust to the prices.

It was, at this time that CECEI, the French regulatory body intervened and decided not to let Société Générale go ahead. They were concerned with the nature of job losses that such a merger would entail and the impact of the merger on the French banking industry. They wanted to create a strong French megabank which would not be susceptible to external bids and would become a global player in the French banking industry.

On July 1, 1999 BNP came up with an improved offer for Société Générale and Paribas. They offered $41 billion for the two banks. This marked the turning point of the battle. It was at this point that the three banks started going over board and the market began to, believe that company ego and not strategic intent was now governing the action of the banks. The market felt that BNP was paying too much. BNP was primarily interested in Société Générale and had to, make a bid for Paribas only because Société Générale did so. The stock of BNP showed a net negative return and so did the stock of Société Générale. The management of the banks then began to woo their shareholders. Société Générale had by this time started feeling that it would lose the battle for Paribas. It was now more interested in keeping its independence. It is not clear how much of share holder value creation was borne in mind while taking these decisions. Société Générale managed to persuade its major shareholders like Axa and Alcatel not to tender their shares to BNP.

BNP was already paying too much for the deal. Société Générale and Paribas however went a step furthur in share holder value destruction. They rejected the new improved BNP offer. The return of the shares of all the three banks fell. It was clear that whatever be the outcome now, there will be no clear winners. Meanwhile, Paribas middle and lower management was getting increasingly disenchanted with the behaviour of Société Générale. Although the top management was with Société Générale the lower levels began to, get sceptical. They did perceive the merger to be one of equals and started supporting the BNP bid. Société Générale was looked at as an aggressor, a bank that probably wanted to, gain complete control over Paribas. Moreover, BNP began to, promise a merger with minimal layoffs.

Société Générale and Paribas probably should have accepted the second BNP offer. For BNP it had now become a prestige issue. The shareholders of BNP were not ready to, pay such a premium. However, the shareholders of Société Générale and Paribas would have gained. The only problem might have been the ego of Société Générale top management. It was probably hard for the top management to accept that BNP was ready to pay more than what the Société

Générale management could have given its shareholders. Moreover, this would have been the right opportunity for Paribas management to, reduce friction with BNP, since it was quite clear that the battle for Paribas was almost over.

By August 13, 1999 BNP had acquired 65% stake in Paribas and approx. 40% stake in Société Générale. At first glance, the market adjusted return of 7% for BNP shares might convey optimistic sentiments and support for the merger, this rise was merely a reflection of relief with the ending of the battle in sight. BNP share prices (market adjusted) had plummeted when it came up with its improved offer and when Société Générale had rejected the offer. This rise did not take the returns anywhere near the original levels.

Société Générale stock soared. Again, one must bear in mind that its share had been showing substantial negative rates of (market adjusted) retum. The soaring of the stock was merely a reflection of the stabilisation sentiment. This could have occurred because of two reasons:

Shareholders who were against the merger with BNP thought that BNP would now have to, withdraw from its attempt to control Société Générale and would have to return the shares

Shareholders who thought that with 40% control BNP might seek the support of CECEI and other shareholders for the rest of Société Générale and that it would succeed

It is difficult to say why the share prices went up but the shareholders of Société Générale like BNP were worse off than when it had started. All the analysis done by these companies tend to portray favourable stock movement but they do not account for the following:

Market adjusted return (accounting for Beta)

The position of the stock compared to, the beginning of the battle. A 20% decline followed by a 7% rise cannot be treated as a situation with net positive value

CECEI and Mr. Trichct of thc bank of France meanwhile began to, realise that anything short of a three-way merger would be a loss for French banking industry. Their aim of a French banking giant would not be realised and Société Générale might become vulnerable to, foreign banks. The BNP Paribas deal was allowed to go through and the regulatory authorities began trying to broker a deal between the banks. Several suggestions were made including holding comparable stakes in each other and the BNP offer of letting Société Générale continue as a separate entity. At this time it was obvious that Société Générale would not allow any sort of control by BNP. It would have lost face. It could probably have donc wcll by holding comparable stakes in BNP. The only plausible reason for the rejection of this seems to be that the much higher bargaining power of BNP shares after the merger and the 38.75% stake that BNP already had in Société Générale. They feared loss of independence, something that had guided their course of action so far. Their fears were probably valid in the light of BNP's intentions. It is however, not clear what would have been good for shareholders at this point.

When repeated negotiations between BNP, Société Générale and CECEI did not yield any results, the final decision was taken. BNP would have to return its holding in Société Générale to share holders. CECEI did not feel that it would be in the interest of Société Générale shareholders to have BNP as a significant minority shareholder.

The course of events after this was fairly obvious. Levi Lang, Paribas chief quit. BNP launched a bid for the rest of Paribas and as of December '99, it has 96% of Paribas share.

The outcome of the voting in which BNP attained control of Paribas but could not do the same for SG has left SG isolated. This will force the latter to seek cross border alliances, an option the French governement clearly wants to discourage. There are speculations that SG might seek alliances with european powerhouses such as Spains leading bank BSCH, Britain's biggest composite company CGU Plc or Dutch banking giant ABN Amro.

Though in this two way merger the strategy of SG was left totally in ruins BNP also could not acheive its real prize- the merger of France's two largest retail banks generating big cost savings and creating a bank capable of playing a leading role in the rationalisation of the industry on a European basis is also defeated.

6 A Comment on the Synergies Expected from the Merger Between BNP and Paribas

6.1 Terms of the offer

29 BNP shares in exchange for 20 Paribas shares

-No CVG (contingency value guarantee) offered

> Premiums offered over share price

> September 17: +7.4%

> March 9: +33.8%

> 1 -month average: +9.3 %

> 3-month average: +9.0%

> 6-month average: +10.2% > 12-month average: +29.9

The total pre tax synergies expected by year 2002 are Euro 850 million. Of this, they expect to gain Euro 150 million in revenue enhancement. This represents 1% of the combined sales of BNP and Paribas in 1998. The rest is expected to be realised through cost cutting. Of this Euro 700 million of cost cutting, savings in Information Technology will represent Euro 250 million.

It is important to note that BNP expects to have 48% of its synergies from Corporate and investment banking. The main focus of activities for BNP is retail banking and this is where the maximum synergies would have been derived on merging with Société Générale.

All these synergies are not expected to occur immediately, BNP-Paribas expect to achieve Euro 170 million (20% of final synergies) in year 2000, and Euro 425 million (50% of final synergies) in year 2001.

The BNP-Paribas combine is quoted at a Price to Book ratio of 1.9x. This is low compared to the average European ratio of 2.7x. This is an indication of the markets lack of confidence in the ability of the merged entity to realise the promised synergies.

BNP thinks that significant cost savings will accrue, but given the lack of overlap between the two businesses, these savings may prove to be elusive.

6.2 Cost Cutting

The cost cutting is primarily expected to, arise for the following reasons:

Merger of corporate banking and investment banking businesses will lead to:

- Merger of back and middle offices and support functions
- Merger of international networks
- Economies of scale in research
- IT rationalisation

6.3 Revenue Enhancement

The combined entity expects to enhance its revenues primarily on the basis of two reasons:

By leveraging its larger size :

Using the complementary nature of its business and larger customer base to, enable cross selling:

The presence of complementary activities gives considerable scope for providing a one stop shop for its clients and also for cross selling of its products to the now larger client base.

6.4 Reallocation of capital

The combined entity will reduce the capital allocated to Paribas capital by one third. The allocation to Corporate and Investinent Banking will also be reduced. BNP Paribas expects to reinvest at least Euro 1200 million in external growth (Euro 700 million in International Retail and Euro 500 million in Special Financial Services). These measures, together with a share buyback program are intended to increase the Return on Equity to 16% by year 2002.

7 Conclusions

The analysis of the operation shows clearly that the two commercial/ universal banks BNP and Société Générale were keen to consolidate their position because they felt that they are likely to be targets for takeover bids given their relative sizes in terms of market capitalisation. The move on their part was more a defence mechanism rather than a proactive approach for value creation for the shareholders. This argument is reinforced by their very low market value to book value ratio; for both these banks it was around 1.7 when the average for the European banks was 2.7. In addition, the ROE of the two banks was well below the expected risk adjusted return. The Beta values of the three banks before the merger operation were as follows:BNP = 1.63; Société Générale (SG) = 3.24; Paribas = 0.56.

Based on our estimates, the risk adjusted returns on BNP should have been 14% and for SG 21%. BNP performance was not from the market expectations, whereas, for SG figures were not upto the expectations. Only Paribas ROE was compatible with the market expectations. Paribas was therefore an attractive target.

The detailed analysis of the operation shows that the synergy gains are difficult to estimate, especially in a sector where the economic environment is changing rapidly, but what really matters, is the not the real gain but the markets perception. The market reaction to the successive announcements confirms the theory of asymmetric information. The decision by BNP to pay by shares indicates that the management of BNP were not convinced of the synergy gains. It appears that the pre merger analysis did not give serious consideration to the post merger implementation problems. The post merger problems are the most important cause of the failure of a merger operation(KPMG report).

In the case of M&A operations the monetary authorities tend to play a decisive role, this is particularly true in France. In this case the Commission for Lending institution (CECEI), the French authority overseeing the activities of the banks and financial institutions in France played an important role in the final decision. In fact, they denied autorisation when SG made the improved bid for Paribas. They also forced the two banks BNP and SG to get rid of their cross equity holdings, as they felt that it would have a negative impact on the future development of the two banks.

Two other factors played an important role, firstly, the personal egos of the CEO's and secondly, the trade-unions. In this particular case, the merger of the three banks was refused because the authorities were scared that it could lead to serious labour problems.

The results of BNP/Paribas in 2000, i.e. after one year of the merger operation seems to be positive. The net income rose by 26.3%, this was significantly higher than forecasted net income. The announcement led to an increase of the share price by 5%. The ROE improved to 20.9%, an increase of 4% over the previous year. The synergy gains over the year amounted to Euro 268 million as against the planned figure of Euro 140 million.

The restructuring in the European banking sector is still continuing. The two British banks, Lloyd's and Abbey National are negotiating merger opportunities. This means that Société générale has to consolidate its position may be through trans border operations, which could include strategic alliances.

References

P.Asquith, R.F. Bruner and D.W. Mullens, " Merger Returns and the form of financing"; Working Paper, MIT Sloan School of Management, August 1990.

Chevalier Alain, Gupta Jyoti, "Opération de concentration: quelques éléments de comparaison Européene", Analyse Financière, No 80, 1990.

J.R.Franks , R.S.Harris, " Shareholder wealth effects of corporate takeovers: the U.K. experience, 1955-1985, Journal of Financial Economics, May 1989.

M.C.Jensen, R.S.Ruback, " The market for corporate control: the scientific evidence", The journal of Financial Economics, April 1983.

W.G.Lewellen, " A pure financial rationale for the conglomerate mergers", Journal of Finance, August 1990.

S.C.Myers, " A framework for evaluating mergers", Mordern developments in Financial Management, S.C.Myers (Ed), Frederick A. Praeger INC, New york 1976.

R.S.Ruback," The cities service takeover: A case study", Journal of Finance, May 1983.

Restructuring of Development Banks in the Context of a Market Economy: The Case of India

D. Singh, N. Dhillon

Management Development Institute
Gurgaon
India

Summary:
The Indian national level institutions in consort with the specialized financial and state level institutions have been important channels in the movement of credit through various product and service offerings to the coprorate sector. With the financial sector reforms and the opening of a market based economy they face the strategic challenge of metamorphosing into universal banks. They have also to confront a newer regulatory environment, strike a new equation with multilateral and regional agencies and balance the changing magnitude and urgency of commercial and developmental priorities. The paper looks at the Indian experience with the reforms of the All India Development Banks, circumstances of their adjustment and their strategy and performance.

Keywords: Reforms, Regulatory changes, Financial performance

1 Introduction

The Industrial Development Bank of India (IDBI), The Industrial Credit and Investment Corporation of India (ICICI) and The Industrial Finance Corporation of India (IFCI) are three major apex level Development Financial Institutions (DFIs) in India. The trio of DFIs has focussed largely on term finance unlike the other All India Financial Institutions, which have been assigned goals other than predominantly term financing. They have played an important role as financial intermediaries and as catalytic agents for the overall economic development of the country during both the pre-reforms as well as the post-reform phase of the country's economic growth. These three institutions are the focus of this study as they typically account for 75% of the total financial assets of the All India Term

Lending Financial Institutions. Further, they are in the midst of adapting to a rapidly changing financial and business environment. The paper looks at the circumstances of their adjustments during the post and pre liberalization periods. What has been the Indian experience with respect to the DFIs and what else needs to be done to accompany the process of financial liberalization? It needs to be emphasized that the financial reforms is an ongoing process and encompasses a wide range of measures besides credit allocation and liberalization.

2 Pre-reform Circumstances

An examination of the post-reform performance of the DFIs must begin with the examination of the pre-reform circumstances of their birth. India opted for a state-planning model at the time of its independence and the state took upon itself the onus for leveling the countries economic problems. The DFIs were viewed as the agents to achieve these goals. Multilateral institutions too, preferred national level institutions through whom they could channelize aid. Credit was channeled to enterprises and sectors depending on their relative importance in the overall schema of self-reliance and a self-sustaining growth. In the process certain sectors of the economy got a larger allocation of the scarce monetary resource at cheaper rates. On the demand side, the need for finance in these sectors was largely driven by capacity expansions. These decisions were a fallout of the licensing regime in place and therefore the DFIs had to primarily ensure that funds were available and given for the purpose of expansion. The DFIs funded projects, were not only deemed crucial to the industrialization of the country but were also expected to bring about significant economic and social externalities. DFIs were responsible for developing regions and supporting industries that were not considered bankable by the traditional banking system. It was but natural that besides the economics of the projects, decision-making criteria also factored in the political and social aspects of the investments. Unlike commercial banks, which accepted deposits for lending, development financial institutions have had a more investment orientation. And further, unlike commercial banks the liability side of their balance sheet has been less of a concern to them than their asset side. A positive fallout of this process has been that over the years they have developed competencies in project appraisal and experienced board level decision making in several of the large and medium enterprises by virtue of their funding through long term loans and equity capital.

3 Rationale Behind Reforms

Reforms were introduced in India to hasten the process of financial deepening and reduce the fragmentation of the financial markets. Typically the DFIs operated in a financial system portrayed as repressive and fragmented. Aryeetey et al (1996) characterize repressive financial policies when there are restrictions on market entry, public ownership of financial institutions, high reserve requirements, interest rate ceilings, quantitative controls on credit allocation and have restrictions on capital transactions with the rest of the world. Repressive financial polices in turn lead to a fragmentation of the financial system. Two analytical models are used to explain the fragmentation of the financial system. Fragmentation arises because there are ceilings on deposits and loan rates and as a result demand for capital goes up not commensurate with the supply of investible funds. Unsatisfied demand for funds leads to rationing or credit allocation by means other than interest rates. As a result there are borrowers who can access funds at very low rates of interest and there are others who seek expensive credit from other markets leading to a fragmentation of the financial market. McKinnon (1973), Shaw (1973), Fry (1982) (1988) and Roe (1991) are the proponents of this model.

An alternate model used to explain the process of financial fragmentation has been propounded by Stiglitz and Weiss (1981) Stiglitz (1989) and Hoff and Stiglitz (1990). According to this model, fragmentation takes place in developing countries because of financial and institutional features of the markets. There are differences across various segments of the financial markets due to imperfect information on credit worthiness, monitoring and screening costs and contract information. These in return result in imperfect or adverse selection and give birth to the moral hazard problem. If interest rates increase in certain segments, borrowers with worthwhile projects are discouraged to seek loans. For a DFI the quality or mix of the loans changes adversely and the borrowers will now seek higher risk projects thereby increasing the risk of default. The moral hazard problem arises when some borrowers take loans to recycle or not pay at all. Thus the level of regulated interest rates affect the composition of the loan portfolio of the DFIs. DFIs when faced with the greater risk may resort to non-price rationing rather than raise the interest rates when faced with excess credit. Thus, for the DFIs credit rationing may characterize market equilibrium even in the absence of interest rate ceilings and direct allocation of credit. Poor information systems and absence of common linkages across markets encourage segmentation. Financial institutions such as the DFIs are faced with higher information acquisition costs and although smaller institutions and informal agents can get this information at a much lower cost they do not have the capacity to scale up. Further weaknesses in the overall financial system such as costs of contract enforcement, absence of an insurance market and market for sale of confiscated collateral and increased risk and costs of default.

Further a fragmented financial system would lead to higher costs as shown by Calomiris (1995). Finance costs for industry were lower in Germany than in the United States, because U.S. regulations prevented the universal banking from which Germany benefited.

These traits characterized the environment in which the three DFIs operated. The markets were fragmented on the basis of time and the nature of activities. Firms seeking long term loans would inevitable find themselves before the DFIs. With regulation of interest rates the default risk could not be entirely factored in the loan pricing. Commercial Banks were loath to enter the medium to long term spectrum because of lack of adequate knowledge and information. The control, regulatory and institutional barriers encouraged fragmentation across segments.

4 Post-reform Circumstances

Changes in the business environment mandated that very soon changes were inevitable in the financial environment. It was inevitable for the DFIs as the circumstances and the financial and business environment in which they had earlier prospered had dramatically changed. Together with the global flow of finance, technological changes and increasing realization by governments that they had abrogated on themselves too many activities, the role of DFIs came under inspection. Also notable was the conflict over space between the commercial banks and the non-banking finance companies. As pointed out by Diamond (1997), ' the simplistic dichotomous opposition of commercial bank versus term financier, trade finance versus investment', was no longer able to satisfy the diverse financial needs. The focus of policy was now more on the overall financial system than on a particular type of financial institution. [Diamond (1997)]. Diamond goes onto identify traits of some of the performing DFIs based on his study of Korea Long Term Credit Bank (KLB), Development Bank of Singapore (DBSBANK), Industrial Credit and Investment Corporation of India (ICICI), Banco Portgues De Investimento, S.A (BPI) and Corporacion Financiera Del Valle, S.A. (CFV). Successful development banks have exhibited the following traits among others:

1. Have meshed with the overall financial system
2. Have managed the liability side of their balance sheet as well as they were doing the investment side
3. Diversified their assets and activities and provided a wider range of financial services

The Indian experience with the reforms of the DFIs has been as follows:

4.1 Restructuring of Ownership and Operational Autonomy

The circumstances changed with the reform period in 1991. It began with restructuring of the ownership and operational autonomy. ICICI was established under the Indian Companies Act and at the time of its creation was a completely private company till nationalization transferred a large part of its ownership to the government. A public issue of equity in 1991 saw state ownership fall to around 36% although in terms of ownership and voting power ICICI remains a quasi government entity. However the government has continued to name only a single member to ICICIs Board of 15 Directors and has not taken advantage of its 'adventitious position'. IFCI was converted into a public limited company paving the way for private ownership. The government control was brought down to about 50% through a public issue of equity in 1993. Restructuring also saw Government control of the largest DFI, IDBI brought down to 74.6% by a public issue of equity as against 100% earlier. IDBI is a government body under a statute, but the amended Act provides for private participation in equity subject to the government holding being retained at not less than 51%.

4.2 Regulatory and Supervisory Changes

There have been important regulatory and supervisory changes for the DFIs. These changes pertain to Licensing, Asset Provisioning Norms, Capital Adequacy Requirements, Exposure Norms and On site and Off site Supervision. IFCI, IDBI and ICICI did not require a license because they were set up as statutory bodies. Under the reform process IFCI is now registered under the Indian Companies Act and is treated as a non banking finance company and is required to obtain a Certificate of Registration from the Central Bank, The Reserve Bank of India (RBI). These licensing requirements are similar to those required of Commercial Banks and Non Banking Finance companies. From April 1, 1993 DFIs have been following Prudential Norms relating to income recognition, asset classification and provisioning on an objective criterion of interest and principal recovery. Assets are now classified into four categories, viz., Standard, Sub-Standard, Doubtful and Loss Assets. All the DFIs are required to maintain 12% risk weighted capital adequacy ratio. This has been gradually increased from 8% in 1992. In terms of the exposure norms, the exposure of the DFIs to a single borrower cannot exceed 25% of its net owned funds and should not exceed 50% for Group Borrowers. These exposure norms are the same as prescribed for commercial banks. Important distinctions between the regulatory and supervisory changes for commercial banks and DFIs relate to reserve requirements and priority sector lending. The DFIs are not required to maintain statutory deposits of cash and government denominated securities and have no prescribed priority sector targets. In terms of supervision the DFIs are brought within the supervisory jurisdiction of the Board for Financial Supervision (BFS). The Board will focus on the DFIs capital adequacy, asset quality, management, earnings and liquidity and systems and controls - a modified version of the CAMELS model. The accompanying rating system for the CAMELS model is yet to be worked out. Off site surveillance system is not yet in place.

4.3 Operational Performance

In the post reform period competitive market conditions, more than anything else, determine the decision of industries to add capacity leading to volatility in the demand for capital. The government does not define priority areas in the way it used to. The requirements of the industry have scaled upwards manifold and DFIs are no longer able to meet the entire financial requirements of a firm. On the supply side of capital the government does no longer provide the low cost, assured long-term funds. Several of the sources of low cost concessional finance and government guaranteed bond facilities were withdrawn in the post reform period. Disintermediation has put pressure on both their financing and investment capabilities. The DFIs combat competition from other banking and non-banking financial institutions. Most importantly they are grappling with debilitating non-performing loans, varying risk exposures and tighter regulatory, provisioning norms and capital adequacy requirements and simultaneously choosing investments that meet the institutional goals.

4.4 DFIs as part of the Overall Financial System

The 3 DFIs appear to have meshed well with the overall financial system during the post liberalization period. A glance at Figure 1 indicates that the DFIs have been able to increase their share of the Aggregate Financial Assets and also in comparison to other All India Term Lending Institutions and the Commercial Banks. Their share of the Aggregate Financial Assets in the pre-reform period was between 9-10%. During the reform period they have held between 10-12% of the Aggregate Financial Assets. The share of the commercial banks with whom they are in direct competition has come down. The DFIs have also logged impressive growth rates during the periods of 1997 and 1998. Figure 2 indicates that their growth slowed down immediately after the reforms got underway during which period the commercial Banks grew at a faster pace. Therefore, the DFIs have been upto the challenge posed by the liberalization of the financial sector. They have not lost their market share and have grown and created assets more than that of their nearest rivals, the commercial banks.

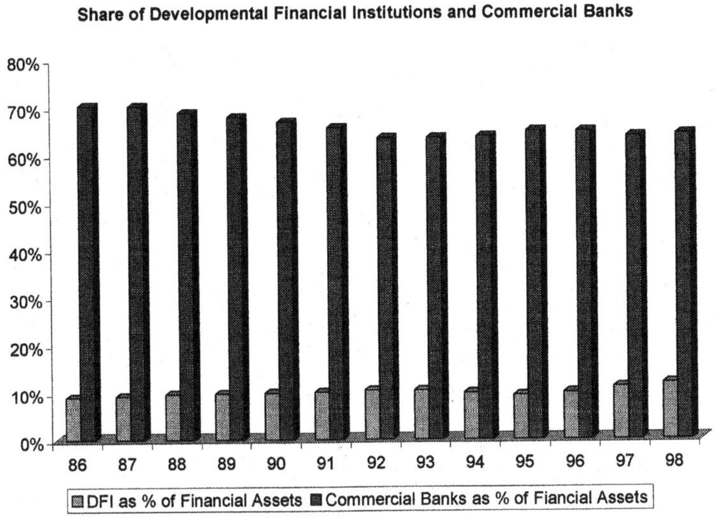

Figure 1: Assets of Financial Institutions and Banks

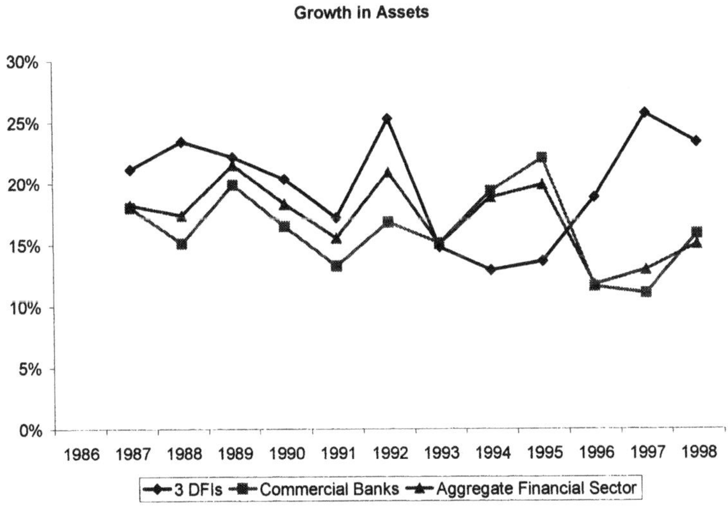

Figure 2: Assets of Financial Institutions and Banks

Source: Data compiled from IDBI Reports on Development Banking

5 Response of the DFIs

The strategic response of the DFIs has been to move away from their traditional segment of medium to long term financing by undertaking short-term lending including working capital finance. They accept short to medium-term deposits and Certificate of Deposits and have resorted to newer and innovative sources of raising loan and equity capital. Both these measures have widened their sources and deployment avenues.

5.1 Diversification of Activities

Another strategic response of the DFIs to the changing context has been their transformation and foray into Universal Banking. Universal Banks as defined by Calomiris (1995) are large banks, with extensive networks of branches, provide many different services, hold several claims on firms (including equity and debt), and participate directly in the corporate governance of firms that rely on the banks for funding or as insurance underwriters. Table 1 indicates their activity portfolio and the extent of diversification into housing, credit rating, merchant banking, venture capital, commercial and investment banking, investor servicing, broking, custodial services, electronic trading in stock exchanges, registrar and transfer services and short term funding through different forms of ownership. Most of these activities were pursued after the reform process. They have resorted to the route of wholly owned, majority holding subsidiaries for most of the above activities. In addition they have contributed to the equity capital of several other organizations engaged in depository, term lending, biotechnology etc. This process of fragmentation and renewed consolidation is similar to the process of integration of giant financial institutions, which are operating as holding companies for a plethora of financial services and products in Europe and the US. The choice of activity and the ownership route taken are different for each of the DFIs. Nevertheless they indicate the strategic response of the DFIs to metamorphosise into Universal Banks

5.2 Diversification Across Industries

The DFIs have shifted their investment activities across different industries. In the pre-reform phase the industries that got most of the sanctions and disbursements experienced declining allocations in the post reform phase. A larger portion of the allocation of investible funds was channeled to the infrastructure sector. This indicates that the DFIs have been active in meeting their commitments to the 'developmental goals'. There was a minor change in the sanctioned and disbursed industry portfolio in the post-reform phase as some industries that accounted for almost 80% of the investments in the pre-reform phase were not in the list for the post-reform phase. Table 2 indicates the percentage allocations across different industries and the percentage changes in the allocation of investible credit during the periods under study. A comparison of the industries that account for almost 80% of the allocation with those of the commercial banks reveal that the industries that both the Commercial banks and DFIs are exposed to are fairly similar. This lends credence to the view that commercial banks tend to follow the industry choice of the DFIs because the cost of information processing, monitoring of loans and other forms of finance may be prohibitive for them. In essence the DFIs fulfill an important 'developmental role' as pioneers and a front desk for finance.

Table 1: Subsidiaries Set up by the Developmental Financial Institutions

Activity	IDBI	ICICI	IFCI
Asset Management		W / M	
Bio-Technology, Tourism, Housing	O	O	O
Brokerage Services		M	
Capital Market Services	W		
Commercial Banking	M	M / O	O
Credit Rating	O	O	O
Custodial Services	O		W / O
Depository	O		
Factoring, Leasing and Financial Services		W	
Infrastructure Financing		M / O	
International Services		W	
Investment Banking	W		
Investor Services	O	W	W
Merchant Banking		M	W
Mutual Fund	M		O
Off shore, Equity funds		W	
Refinancing to Small Scale Sector	W		
Securities Exchange	O		
Term Lending	O		
Trading in Money Market Instruments, Securities	O	M / O	O
Venture Capital		W	M

Source: Compiled from RBI Documents

W: Wholly Owned Subsidiary/Affiliate

M: Majority Owned Subsidiary/Affiliate

O: Other Affiliate

5.3 Increase Allocations for the Private Sector

An indication of the increasing independence of the DFIs in investment decision making has been their higher allocations for the private sector. Before the reform process got underway, the private sector accounted for 78% and 76%of the sanctions and disbursements respectively. In the period 1993-99 the private sector accounted for 84-85% of the sanctions and disbursements respectively. The share of the government allocations has come down despite it having a higher growth as compared to that of the private sector. Table 3 shows the allocation of investible funds across different sectors of the economy.

Table 2: Industry wide allocation of Investible Credit by the DFIs

	Percentage (%)			Change p.a.		
Sanctions/Disbursements	1986-92	1993-99	1986-99	1986-92	1993-99	1986-99
Food Products -Sugar and Others	5%	3%	3%	18%	10%	16%
Textiles	11%	8%	9%	22%	9%	17%
Paper and Paper Products	2%	2%	2%	30%	15%	21%
Rubber and Rubber Products	2%	1%	1%	16%	12%	21%
Chemicals and Chemical Products	14%	12%	12%	19%	6%	14%
Crude and Petroleum Refining	0%	6%	5%		66%	
Fertilizers	5%	4%	4%	6%	22%	19%
Cement	5%	4%	4%	13%	11%	14%
Basic Metals	9%	9%	9%	22%	8%	21%
Metal Products	2%	3%	3%	10%	14%	14%
Machinery	4%	3%	3%	21%	14%	18%
Electrical Machinery	3%	2%	3%	26%	18%	19%
Electronic Equipment	3%	3%	3%	37%	17%	28%
Transport Equipment	3%	3%	3%	17%	24%	23%
Infrastructure	9%	23%	20%	12%	48%	29%
Services	11%	8%	9%	3%	27%	19%
Other Industries	11%	6%	7%	4%	11%	10%
Total	100%	100%	100%	15%	22%	21%
Disbursements						
Food Products -Sugar and Others	6%	4%	4%	17%	7%	14%
Textiles	11%	10%	10%	20%	16%	19%
Paper and Paper Products	2%	2%	2%	10%	22%	16%
Rubber and Rubber Products	2%	1%	1%	27%	6%	19%
Chemicals and Chemical Products	14%	14%	14%	22%	11%	17%
Crude and Petroleum Refining	0%	5%	4%		143%	
Fertilizers	5%	4%	5%	15%	21%	17%
Cement	5%	5%	5%	-3%	4%	10%
Basic Metals	8%	10%	9%	25%	15%	23%
Metal Products	3%	4%	3%	14%	18%	16%
Machinery	3%	3%	3%	23%	10%	17%
Electrical Machinery	3%	2%	2%	30%	18%	20%
Electronic Equipment	3%	4%	4%	35%	16%	28%
Transport Equipment	3%	4%	4%	25%	23%	24%
Infrastructure	9%	13%	13%	9%	36%	23%
Services	13%	8%	9%	3%	25%	16%
Other Industries	10%	7%	8%	10%	5%	11%
Total	100%	100%	100%	15%	19%	19%

Source: Data compiled from IDBI Reports on Development Banking

5.4 Increase in Allocations for Working Capital Purposes

Traditionally, DFIs have been allocating their investible funds for the purposes of new investments, modernization and expansion. These end uses are largely of a capital nature and accrue towards the fixed capital of the company. During the days of the state planning, 91% of the DFIs investments were in these areas.

However, during the post reform phase the allocation of investible funds is down to 75%. Working capital loans have begun to account for a larger share of the allocable funds as shown in Table 4. This is an indication that the DFIs are migrating to another turf because during the period of 1986-92, loans for working capital purposes fell principally within the domain of the Commercial Banks. It could also mean that the Commercial Banks too have entered project finance and caught into the share of the DFIs. Nevertheless it is an indication of the degree of competition appearing on the financial scene in India and the disappearing boundaries between financial institutions.

Table 3: Allocation of Investible Funds Across Different Sectors

Sector Wise	1986-92	1993-99	1986-99	Growth p.a.		
				1986-92	1993-99	1986-99
Sanctions						
Public	13%	11%	11%	9%	36%	20%
Joint	7%	4%	5%	22%	16%	17%
Cooperative	2%	1%	1%	4%	15%	5%
Private	78%	84%	83%	16%	21%	22%
Total	100%	100%	100%	15%	22%	21%
Disbursements						
Public	14%	9%	10%	11%	27%	17%
Joint	7%	5%	5%	13%	25%	18%
Cooperative	3%	1%	1%	-2%	-20%	-12%
Private	76%	85%	83%	17%	18%	20%
Total	100%	100%	100%	15%	19%	19%

Source: Compiled from RBI Documents and IDBI Report on Development Banking

Table 4: Allocation of Investible Funds Based on End Use

End Use	1986-92	1993-99	1986-99	Growth rate p.a.		
				1986-92	1993-99	1986-99
Sanctions						
New	40%	36%	36%	14%	26%	23%
Expansion and Diversification	27%	28%	28%	37%	15%	29%
Modernization/Balancing/Equipment/	24%	11%	13%	35%	7%	20%
Rehabilitation	0%	0%	0%	81%	-14%	27%
Others incl. Working capital loans	9%	25%	23%	24%	43%	43%
Total	100%	100%	100%	25%	24%	28%

Source: Compiled from RBI Documents and IDBI Report on Development Banking

5.5 Form of Investments

DFIs have traditionally been giving direct assistance primarily in the form of Rupee loans and short-term finance for bridging capital requirements. Assistance in the form of Guarantees has not been significant. During the period 1986-92, Direct Assistance in the form of Rupee and Foreign Currency loans, Direct

Subscription to Equity, Preference Shares and Debentures and equipment leasing and technical assistance together accounted for about 70% of the total investible funds sanctioned. The balance was in the form of Non Project finance and Short term loans. Following liberalization the reform period saw that Direct Assistance has shot upto 86-90%. DFIs were more active in direct participation to the equity and debt issues of firms. There was more involvement of Rupee and foreign currency loans and short-term bridge finance and corporate loans were lower as compared to the previous period. Leasing and Hire Purchase, traditionally the domains of the Non Banking Finance Companies were not areas that the DFIs either enlarged or reduced their activities. Table 5 tabulates the forms of assistance given by the 3 DFIs during the period under study.

Table 5: Form of Investible Funds Sanctioned and Disbursed

Sanctions	1986-92	1993-99	1986-99	Growth rate p.a. 1986-92	Growth rate p.a. 1993-99	Growth rate p.a. 1986-99
Asset Creation	69%	89%	85%	27%	23%	26%
Direct Finance/Assistance	60%	73%	70%	22%	28%	26%
Rupee loans	43%	45%	45%	21%	30%	26%
Foreign currency loans	9%	12%	11%	21%	18%	20%
Direct Subscription of Equity/Preference shares	2%	4%	4%	10%	30%	30%
Direct Subscription of Debentures	0%	6%	5%	-100%	51%	82%
Deferred Credit/Direct discounting of Bills	4%	3%	3%	36%	6%	20%
Equipment Leasing/Hire Purchase/Technical Development	3%	3%	3%	18%	15%	24%
Underwriting Guarantees	4%	13%	11%	35%	13%	31%
Underwriting of Equity and Preference shares	3%	2%	2%	34%	-39%	-3%
Underwriting of Debentures	0%	1%	0%	-100%	-41%	-4%
Guarantees	1%	10%	9%	40%	30%	39%
Assistance to Leasing companies / Refinance / Rediscounting / Fin services	4%	4%	4%	206%	-15%	60%
Non Project Finance/ Short term corporate loans	31%	11%	15%	-11%	15%	8%
Total Assistance Sanctioned	100%	100%	100%	16%	22%	21%
Disbursements						
Asset Creation	65%	86%	82%	25%	19%	23%
Direct Finance/Assistance	59%	79%	75%	22%	22%	23%
Rupee loans	45%	49%	48%	20%	22%	22%
Foreign currency loans	8%	14%	13%	34%	11%	22%
Direct Subscription of Equity/Preference shares	1%	4%	3%	31%	36%	41%
Direct Subscription of Debentures	0%	6%	5%	n.a.	65%	n.a.
Deferred Credit/Direct discounting of Bills	3%	2%	2%	24%	15%	19%

Table 5: Form of Investible Funds Sanctioned and Disbursed (continued)

Sanctions	1986-92	1993-99	1986-99	Growth rate p.a. 1986-92	Growth rate p.a. 1993-99	Growth rate p.a. 1986-99
Equipment Leasing/Hire Purchase/Technical Development	2%	3%	3%	17%	21%	26%
Underwriting Guarantees	2%	2%	2%	-10%	41%	12%
Underwriting of Equity and Preference shares	1%	0%	0%	11%	3%	12%
Underwriting of Debentures	0%	0%	0%	n.a.	12%	n.a.
Guarantees	1%	1%	1%	-13%	61%	12%
Assistance to Leasing companies / Refinance / Rediscounting / Fin services	5%	5%	5%	n.a.	-16%	n.a.
Non Project Finance/ Short term corporate loans	35%	14%	18%	-11%	15%	8%
Total Assistance Disbursed	100%	100%	100%	15%	18%	18%

Source: Compiled from RBI Documents and IDBI Report on Development Banking

5.6 Return Performance of the DFIs

The performance of the DFIs has come down from the level of 1992-93. The Return to the Equity shareholders has fallen steeply from the level of 17.5% to 11.2%. The DFIs have been under pressure from not only the competition from other financial institutions but also from the regulators. The spreads have been under pressure because of declining interest rates and their inability to source low cost capital. Simultaneously, the leverage enjoyed by them before the process of liberalization started is under scrutiny as they adhere to newer capital adequacy requirements. There has been a marginal improvement in their utilization of the assets but overall the performance has dipped on the return measures. The results for the period 1993-99 are summarized in Table 6. The level of Non- Performing Loan portfolio of the DFIs is now at 12.2 % of the total loan assets. A strict comparison with the earlier period is not possible because of differing loan provisioning norms.

6 Conclusions

The Indian Experience with the DFI indicates three distinct phases in Restructuring of DFIs. There has been dilution in ownership, greater operational autonomy and increased independent supervisory control and reporting systems. The DFIs have pitted themselves directly against the Indian Commercial Banks as indicated by higher allocations for working capital and have held held their ground by creating more financial assets. There are also some positive externalities for the

Commercial Banks as they happen to share similar allocations for the same industries. There has been no direct and significant competition with the Non Banking Finance Companies because of the insignificant allocation towards leasing and hire purchase. As regards the developmental goal there has been a discernible shift in industry mix towards longer gestation industries and infrastructure. The share of the private enterprise sector has grown. There has been greater involvement with companies beyond loans through direct subscription to their debt and Equity capital. The return performance has been declining with increasing pressure on net interest margins. Fund based business is under pressure from Commercial Banks and increasing level of dis-intermediation. There has been support to aggregate returns from fee based activities. Higher provisioning norms have impacted on leverage and therefore there are declining book and market returns to the shareholders.

Table 6: Financial Performance of the DFIs

	1992-93	1992-93	1993-94	1994-95	1995-96	1996-97	1997-98	1998-99	1999-2000
Net Interest Margin	0.53%	0.38%	1.39%	1.93%	2.15%	1.70%	1.74%	0.90%	0.26%
Interest Income	3,589	4,629	5,485	6,711	8,336	10,780	12,951	14,899	15,637
Interest Expended	3,337	4,422	4,638	5,385	6,577	9,007	10,737	13,532	15,218
Interest Spread	252	207	847	1,325	1,759	1,773	2,214	1,367	419
Non Interest Income	519	638	192	137	(60)	301	(40)	559	1,586
Other Income	948	1,216	1,203	1,349	1,707	2,423	2,637	3,373	4,562
Other Expenses	191	282	443	583	954	1,016	1,095	1,334	1,526
Taxes	161	226	288	359	527	591	480	171	203
Loan Loss Provisions	77	69	280	270	285	516	1,103	1,311	1,247
Net Profit	771	845	1,039	1,462	1,700	2,074	2,173	1,926	2,005
Non Interest Income to Total Assets	1.10%	1.19%	0.32%	0.20%	-0.07%	0.29%	-0.03%	0.37%	0.98%
Net Profit to Total Assets	1.64%	1.57%	1.71%	2.13%	2.08%	1.99%	1.71%	1.27%	1.24%
Return on Equity	19.3%	17.5%	17.6%	20.2%	16.5%	15.7%	14.9%	12.5%	10.7%
Profit Margin	17%	14%	16%	18%	17%	16%	14%	11%	10%
Operational Efficiency	10%	11%	11%	12%	12%	13%	12%	12%	13%
Leverage = TA/NW	11.83	11.13	10.28	9.48	7.91	7.89	8.70	9.82	8.59

References

The word "processed" describes informally reproduced works that may not be commonly available through library systems.

Aryeetey, E, H Hettige, M Nissanke and W Steel. (1996). Financial Market Fragmentation and Reforms in Ghana, Malawi, Nigeria and Tanzania. *World Bank Economic Review*, Vol. 11, No. 2, pp.195-218.

Calomiris Charles W (1995). Universal Banking and the Financing of Industrial Development, *Reforming Finance: Some Lessons from History, edited by Gerard Caprio, Jr. and Dimitri Vittas. World Bank*

Diamond, W (1997) Successful 'Development Banks', *International Journal of Development Banking*, Vol. 15, No.2 pp.63-77.

Fry, M.J.(1982). Models of Financially Repressed Developed Economies, *World Development*, Vol.10, No.9, pp.731-750.

Fry, M.J.(1988). *Money, Interest and Banking in Economic Development*. John Hopkins University Press, Baltimore.

Hoff, K and J.E Stiglitz (1990). Imperfect Information and Rural Credit Markets: Puzzles and Policy Perspectives, *The World Bank Economic Review*, Vol. 4 (3), pp. 235-250.

IDBI Reports on Development Banking 1993-98

McKinnon, R I (1973). *'Money and Capital in Economic Development'*, Brookings Institutions, Washington D.C.

Reserve Bank of India Bulletin 1992-98

Reserve Bank of India Reports on Currency and Finance 1992-98

Roe, A.R (1991) "Financial System and Development in Africa." *A Conference Report Of An Economic Development Institutional Policy Seminar, Nairobi 1990 World Bank, Washington D.C (Processed)*

Shaw, E S (1973). *'Financial Deepening in Economic Development'*, Oxford University Press, New York

Stiglitz, J.E (1989). Financial Markets and Development, *Oxford Review of Economics and Policy*, Vol 5 (4). pp. 55-66.

Stiglitz, J.E and A W Weiss (1981). Credit Rationing in Markets with Incomplete Information". *American Economic Review*, Vol 71 (3). Pp. 293-41

Net Interest Income, Balance Sheet Structure and Interest Rates

Ch. Staikouras[1], **G. Wood**[2]

[1] EFG Eurobank Ergasias
Strategy Investment and Planning Division
8 Othonos Street
10557, Syndagma Square
Athens, Greece

[2] City University Business School
Frobisher Crescent
Barbican Centre
London EC2Y 8HB, UK

Summary:
The gap position gauges a bank's exposure to interest rate risk. In fact, however, a number of factors may affect the yields that banks earn and pay on assets and liabilities and therefore their interest income flows. These factors include market demand characteristics, market supply conditions including market structure, and macroeconomic indicators. Regression models are tested to determine the effects of changes in the level and volatility of market interest rates, the funds gap ratio, and market conditions on net interest margin. From the analysis it seems that there is a direct (level and variability) or indirect (through the balance sheet structure) influence of nominal interest rates on net interest margin of the banks. In the long run, the direct effects have the major impact, while in the short run the indirect effects do.

Keywords: Net interest margin, Interest rates, European banking industry, Concentration, Stability

1 Introduction

The sensitivity of a commercial bank's income sources and market values to changes in nominal interest rates is of interest to bankers, regulators, investors, and researchers. Conventional wisdom has long viewed financial intermediaries, such as banks, as being susceptible to interest rate risk. Since the early 1970s, bankers have become increasingly aware of the effects volatile interest rates have on net interest margin. This arises because of banks' propensity to undertake asset transformation or intermediation functions by lending long (often at fixed rates) and borrowing short (often at variable rates). Although there is no generally accepted view as to why 'banks' have an advantage in undertaking this asset transformation function (as compared to other firms) banks typically keep title to the relatively long-term loan portfolios they originate while selling (deposit) contracts of short maturity. A priori, such mismatched asset-liability structures make bank earnings (or stock values) more susceptible to interest rate changes than firms who choose not to specialise in providing asset transformation services.

Interest sensitivity (or gap management) is the popular concept used for managing banks' net interest income and exposure to interest rate risk. The sensitivity or gap position, defined as the relationship between rate-sensitive assets and rate-sensitive liabilities, gauges a bank's exposure to interest rate risk. Assets, liabilities and equity are valued in book terms. Common equity is usually treated similarly to non-sensitive liabilities. Most of the banks have a positive funds gap i.e. the interest sensitive assets exceed the interest bearing liabilities (the same occurs with the sample of our analysis). Nowadays, banks make increasing use of non-deposit sources of funds and began challenging more of these funds into interest earning assets other than loans.

A common fallacy of interest sensitivity analysis is to attribute all credit cycle variation in interest margin to changes in market interest rates. The traditional statistical cost model implicitly assumes that all banks face identical interest rates on various asset and liability items, so that interbank variations in portfolio mix simply reflect different portfolio preferences. In fact, however, a number of factors may affect the yields that banks earn and pay on assets and liabilities respectively and therefore their income flows. These factors include market demand characteristics, market supply conditions including market structure and the cost of non-financial factor inputs, and macroeconomic conditions. Ho and Saunders (1981) found that the size of the interest spread was a function of four variables: the degree of managerial risk aversion, the average transaction size, the competition within the bank's market, and the variability of interest rates. Exclusion of relevant influences from the model will result in inefficient estimation and may lead to biased coefficient estimates.

While profitability of financial institutions has been studied and debated at length, much less has been done in understanding the income sources of the banks. This study differs from the earlier ones in a number of important ways. First, we examine the effects of macroeconomic conditions, balance sheet structure and market conditions on net interest margin of the banks. Also, we introduce in the analysis not only the influence of the level of interest rates, but also their variability. Third, our sample is much larger than previous studies covering a recent time period. By using Fitch-IBCA Bankscope database, we are able to develop a much richer and more informative data set than earlier studies. Finally, we include the lagged values of the independent variables delays in the decision process.

In Section 2 of the paper we present a literature review covering the impact of market interest rates not only on bank net interest income but also on their profit levels. Section 3 and Section 4 are concentrated on the model specification and the data collection process. In Section 5 we evaluate the size, macroeconomic and market structure effects on net interest margin of the EU banks. Finally, in Section 6 we present some concluding remarks arising from the paper.

2 Literature Review

The EU banking systems are facing major changes in the form of increased competition, concentration and restructuring. Banks' responses to the changing financial systems have been most clearly visible in the financial accounts. Pressure on intermediation margins has led banks to offset the decrease in their interest income by shifting the other sources of income such as fees and commissions (ECB, 2000). A number of hypotheses can be advanced about the causes of this decline in net interest income. One explanation frequently put forward is increased competition in both deposit and lending markets. A related hypothesis explaining declines in net interest margin is deregulation, financial innovation and assets' composition. If secure components of the loan book, such as mortgage lending, grow relatively fast then this can lead to a decline of net interest income as a share of total assets. Similarly, increases of interbank lending or of holdings of marketable securities could reduce interest margin.

The impact of market interest rates on commercial bank revenue, cost and profitability has increasingly concerned economists and policymakers as financial market conditions have become more volatile in recent years. Two distinct aspects of market rate conditions may affect bank profits: the *level* of interest rates and the *variability* of rates around their average level within each period. As nominal interest rates decline, the "endowment" income on interest-free liabilities i.e. the income arising because interest bearing assets exceed interest bearing liabilities, is reduced.

A substantial proportion of banking activities involves small investors and borrowers. Differential information and transaction costs in investors' portfolio allocations make them unable to respond fully to a change in the level of market rates. Because of such retail customers, bank profit margins can vary with the level of market rates. Thistle, McLeod and Conrads (1989) have found that (a) balance sheet composition depends on both the level and the change in interest rates, and (b) banks' response to changes in interest rates is different, depending on whether rates are rising or falling (also Howells and Bain, 1998).

Because the numerator components of net interest margin are both interest sensitive, net interest margin does not vary much over an interest-rate cycle. However, since bankers like to price loans and deposits in their own self-interest, which should be aligned with shareholder interests, Sinkey (1998) observes the following movements in bank loan and deposit rates: during an upswing in interest rates, loan rates rise faster than deposit rates; during a downswing in rates, deposit rates drop faster than loan rates. Of course, variable-rate instruments that are tied to some market rate of interest, such as T-bill rate or LIBOR, will move with market rates.

Banks experience short-term profit fluctuations when market rates change, perhaps because they are actively speculating on nominal rate movements. The potential for adverse effects on bank profitability of increased rate variability was noted long ago. Contrary to the conventional wisdom, Flannery (1981, 1983) found that for bank organisations the reported profits generally fluctuate little when market rates change. Individual banks certainly can choose to undertake interest rate risk, but most financial institutions possess a sufficient range of asset and liability choices to hedge their annual profit margins effectively. As he writes: "*... for most banks they* [the fluctuations of interest rates] *represent no serious threat to long-run viability and profits*". Flannery examined the sensitivity of large and small banks' net current operating earnings to interest rate movements. He found a significant long-term impact for only two of the fifteen large banks sampled, while for small banks only seven out of sixty were significantly affected (all in a positive direction). Flannery supports the view that a ceteris paribus increase (for example) in market rate variability has two potential effects on financial intermediary firms. First, firms with a comparative advantage in securities trading would tend to benefit at the expense of less informed traders in the market. Such bank profits are reported separately. Secondly, the public's demand for intermediated securities -deposits and/or loans- might depend on the degree of uncertainty in primary security markets. For example on the loan side, the demand for fixed rate intermediary loans might be expected to increase with higher market rate variability. Elsewhere, Maisel and Jacobson (1978) arrived at basically the same conclusions.

For Scott (1966) bank profitability does not depend to any significant extent upon the level or the pattern of interest rates. For him, there are issues involving the capital values of bond holdings, and the special tax advantages accruing to banks from their ability to offset capital losses against ordinary income that may depend upon the movements of interest rates as well as the levels of interest rates. But these are short-term considerations, and bankers at large should perhaps not be greatly concerned whether monetary policy is 'tight' or 'easy' from the narrow point of view of their concern over firm profitability, although, of course, they should be concerned about monetary policy from the broader point of view of its impact upon economic stability.

The level of market interest rates is used in bank performance studies as proxy for capital scarcity. When capital is scarce, the cost of money is high, something that will be reflected in market interest rates. High interest rates can produce deteriorating asset quality. These high rates contribute to lending problems because some borrowers lack the ability either to pass along or to absorb increased funding costs. However, Samuelson (1945) wrote that *"...the banking system as a whole is immeasurably helped rather than hindered by an increase in interest rates...and commercial banks would profit more than savings banks*". The nominal interest rate increasingly influences the net interest revenues and these in turn influence positively the ratio of net profits over the assets. Several studies give support on the above statement, such as Bourke (1989) and Molyneux and Thornton (1992).

For Revell (1980) banks and other credit institutions will benefit from a rise in interest rates if the proportion of assets on which they can raise the rates is greater than the proportion of deposits on which they must pay increased rates. Santoni (1986) supports the view that an increase in interest rates reduces the bank's capital. In other words, an interest rate change affects the payment stream obligated by the bank's liabilities before it affects the bank's receipt stream, as maturity dates of a bank's assets generally extend beyond those of its liabilities. Consequently, an increase in interest rate levels reduces the expected net stream of dollar receipts as the bank's creditors renegotiate for the higher interest rate, while the interest rate earn by the bank on the existing loans is locked up. Of course, the loans eventually mature and are renegotiated at the higher nominal rate, but the bank's capital is reduced nonetheless.

Anderson (1979) found that a reduction in interest rates cuts the income of banks significantly, while similarly, the high interest rates in the United States during 1980 in US led commercial banks to experience relatively strong profitability (also Opper, 1981). Cole (1981) mentions that the improved profitability of small banks in the period 1977-1980 occurred due to the increase in gross interest earned on assets. High average interest rates provided an opportunity to increase yields on interest-earning assets. Small banks were able to take advantage of this opportunity because of their relatively large holdings of assets with moderately short-term maturities. Hancock (1985) in an examination of eighteen New York-New Jersey member banks for the period 1973-1978 concluded that banks appear to have profits that increase with interest rates. Davis et.al. (1993) note that those banks that reside in a high interest rate environment generally make higher profits than those who do not cover the higher opportunity cost of their own capital.

3 Empirical Implementation

The analysis will be based on cross-country time-series data (panel data). A panel data set offers a certain number of advantages over traditional cross section or time series data sets. The most obvious advantage is that the number of observations is typically much larger in panel data sets. A related advantage is that these data sets may alleviate the problem of multicollinearity. When the explanatory variables vary in two dimensions they are less likely to be highly correlated. A third advantage is that they make it possible to identify and measure effects that are simply not detectable in cross section or time series data. It is sometimes argued that cross section data reflect long-term behavior, while time series data emphasise short-run effects. By combining the two sorts of information a more general and comprehensive dynamic structure can be formulated and estimated.

Many models and corresponding estimation techniques can be used to pool time-series and cross-sectional data. These models differ according to the assumptions about the constancy of coefficients and the various error terms. These models are the Ordinary Regression Model, the Individual Regression Model, the Seemingly Unrelated Regression (SUR Model), the Error Component Model, the Fixed Effects Model, and the Random Coefficients Model.

When N is very large and T relatively small, as in many panel data problems, there is a serious lack of degrees of freedom necessary for the implementation of the individual and the SUR models. This problem is also applied in our case, and this is the reason we prefer to use the ordinary least squares (OLS) and the fixed and random effects models.

The salient distinction between the fixed and random effects models is whether the time-invariant effects are correlated with the regressors or not. Many authors have observed that even when the random effects model is valid, the fixed effects estimator produces consistent estimates of the identifiable parameters. It would appear therefore that, in general, the fixed effects estimator is to be preferred to the random effects estimator unless we can be sure that we can measure all the time-invariant factors possibly correlated with the other regressors. Furthermore, the Hausman test bears on the question of which estimator, random or fixed effects, is to be preferred. The null hypothesis in this instance is that the individual effects are uncorrelated with the included variables. Under the null hypothesis of no correlation between the error and the regressors, the random effects model is applicable.

4 Data

The data have been collected from Fitch-IBCA Bankscope database; the later covers the period 1992-1999. We include data for banks across all European Union countries, except for Luxembourg since comparable figures on interest rates are not available (from Datastream). Initially the sample was unbalanced, however, in our analysis, we have avoided including years 1992, 1993, and 1999 since the availability of data for these years is small. We conclude with a balanced data set i.e. observations are available for all the variables and units-banks at all dates in the period 1994-1998 (5 years time period). We include commercial banks, savings banks, mortgage banks and co-operative banks. Also we include in each country all financial institutions, even foreign banks, to capture the effects of changes in the macroeconomic and market environment of the host country. We use unconsolidated data, only where these are not available do we use consolidated data. We exclude data that seem peculiar e.g. when the proportion of interest bearing assets to total assets exceeds 1 (or 100%), or the figures for the proportion of interest bearing assets and liabilities is very small e.g. 0.12 (or 12%), or large differences exist between interest bearing assets and interest bearing liabilities. However, these cases are very few. So, the whole sample consists of 2,324 banks.

Table 1 shows the number of banks that are included from each EU country in our sample and presents the several type categories of financial institutions.

Within Europe there is considerable heterogeneity amongst the number and type of financial institutions. Germany is the most fragmented market, followed by Italy and France. In most countries institutional separation between commercial banks, savings banks, and other types of depository intermediary continued to prevail. Savings banks is an important sub-sector in Spain and Germany (Spanish savings institutions held about 44% of total bank deposits, the highest proportion of any European Union country). Large number of co-operative banks exist in France and Italy. In other countries institutional demarcations matter somewhat less than in the above mentioned countries. At the opposite extreme is the UK, where former distinctions between clearing banks and merchant banks are no longer operative, and where conversion of mutual building societies to private owned banks has meant that only a rump of mutual institutions remains.

The net interest margin, i.e. the net interest income divided by average total assets, is a measure of portfolio returns before considering the net burden; the latter refers to non-interest expenses minus non-interest income divided by total assets. It is the overall interest yield on the intermediation process. The ratio of rate-sensitive assets minus the rate-sensitive liabilities divided by total assets is called the funds gap ratio. In our regression models we multiple the funds gap by the level of nominal interest rates of the country where the bank is operating.

The level and volatility of interest rates are assumed to be determined in the loanable funds market and, thus, are exogenous to the model. The traditional statistical cost model implicitly assumes that all banks in a country face identical interest rates on various asset and liability items, so that interbank variations in portfolio mix simply reflect different portfolio preferences (Kwast and Rose, 1982). Relevant money market rates are summarised by weekly observations on the interbank three months offered rates. The rate level for each year (measured in percentage points) was the simple average of 52 weekly rates. The source for interest rates is Datastream. Unfortunately this database does not provide data for short-term treasury bills for all the European countries in order to use a second measure for the level of interest rates. The variability measure constructed for each interest rate is the standard deviation of weekly rates around their annual average.

Table 1: Sample of financial institutions by country and type of institution

Country	*Number of banks*	*Country*	*Number of banks*
Austria	31	Ireland	9
Belgium	28	Italy	363
Denmark	67	Netherlands	28
Finland	6	Portugal	25
France	212	Spain	96
Germany	1,327	Sweden	10
Greece	15	UK	107

Type of institution	*Number of banks*
Commercial banks	603
Savings banks	676
Cooperative banks	936
Mortgage banks	109

Source: Own calculations from Fitch-IBCA Bankscope database

We calculate the five-firm concentration index to measure the level of competition in each country, which is simply the cumulative market share of the five largest banks in a country in terms of total assets (Table 2). We take the natural logarithm of the total assets of each financial institution to measure the size effects on net interest margin. The log of assets is used instead of assets to reduce the scale effects. It would seem natural to include in the regression equation a full set of time dummies in order to allow for time effects. However, since the number of years is small (only 4 years) we avoid using time dummy variables.

Table 2: Five-firm concentration in European Union countries (1993-1998) (by total assets)

	1994	1995	1996	1997	1998
Finland	96.67	97.80	96.69	90.90	91.51
Netherlands	85.63	83.51	82.92	86.90	89.80
Ireland	84.44	78.71	72.59	72.77	71.46
Greece	91.98	91.61	89.37	79.33	80.17
Sweden	73.31	72.56	67.27	73.50	75.46
Belgium	70.41	71.11	71.15	67.60	82.68
Denmark	80.02	77.68	78.35	79.01	78.65
Portugal	59.61	62.54	59.79	58.12	62.05
Austria	80.14	58.11	62.07	60.46	60.21
UK	51.54	48.19	43.31	49.21	47.99
Spain	56.84	56.34	55.86	57.78	59.58
France	41.96	41.33	44.97	43.42	44.03
Italy	37.96	36.26	39.64	37.75	47.04
Germany	28.35	29.08	29.96	35.38	39.31

Source: IBCA database and own calculations.

5 Results

The initial specification of the model is

$$m_{jt} = a \cdot (i_{kt} \cdot [a_{jt} - l_{jt}]) + \varepsilon_{jt}$$

where j enumerates the jth bank within country k; t refers to the examined year; m is the net interest margin; a_j the proportion of total assets that are interest bearing; l_j the proportion of total liabilities that are interest bearing; i the market level of nominal interest rates.

Table **3** provides the descriptive statistics for the selected variables by using the balanced sample. From this analysis we can mention the large variation of the level of nominal interest rates and concentration among the EU countries. The mean value of net interest income is 3.07% of average total assets, while that of the funds gap (already multiplied by the level of interest rates of the country) is 0.21%. It must be mentioned that there are some financial institutions that have negative funds gap ratio i.e. the interest sensitive liabilities exceed the interest sensitive assets. It seems that only the size variable is platykurtic and the distribution is flat relative to the normal, while the others are leptokurtic. Also, all the variables have long right tails, with the only exception of the funds gap ratio which has negative skewness and a long left tail.

Table 3: Descriptive statistics for the variables

	mean	*st. dev.*	*min*	*max*	*var*	*skewn.*	*kurt.*
nim	3.0669	1.4392	-2.3345	18.3610	2.0714	3.0816	23.1027
int. rat.	5.0977	2.1639	3.0753	27.2734	4.6825	2.4798	14.9128
funds gap	0.2054	0.3413	-5.7767	2.7614	0.1165	-1.3636	30.4845
conc.	30.8991	11.1563	23.9400	88.7000	124.46	2.5983	6.4563
lnassets	13.5520	1.6051	8.9899	20.0383	2.5766	0.6789	1.0159

Table 4 provides data on the correlation matrix of the variables. A check of the correlation matrix revealed no evidence of multicollinearity for all pairs of variables. There is quite high correlation of the level of interest rates with their variability (as to be expected) and the funds gap (also expected since gap-to-assets ratio is multiplied by the level of interest rates), and a negative correlation of the natural logarithm of assets with the net interest margin.

Table 4: Correlation matrix

	nim	*in. rat.*	*var. of int. rat.*	*funds gap*	*conc.*	*lnassets*
nim	1.0000					
in. rat.	0.2730	1.0000				
var. of int. rat.	0.0479	0.5568	1.0000			
funds gap	-0.1915	0.3262	0.0269	1.0000		
conc.	-0.0223	0.1971	0.2012	0.1400	1.0000	
lnassets	-0.4006	-0.0618	0.0054	-0.0431	0.1415	1.0000

Initially, we run the ordinary least squares (OLS) and within (fixed effects) estimates when the independent variable is only the funds gap. However we present only the fixed effects models since, from the results, these provide the best estimates (

Table 5). In many applications, the easiest way to implement a fixed effects estimator with conventional software is to include a different dummy variable for each individual unit. This method is often called the least-squares dummy variable (LQDV) method. If N is very large, however, it may be computationally prohibitive to compute coefficients for each cross-section unit. In that case, another way to implement a fixed effects estimator is as follows: a) transform all the variables by subtracting bank-specific means, and b) run OLS on the transformed variables. This approach will work perfectly, apart from the fact that the standard errors need to be corrected. The fact that the fixed effects estimator can be interpreted as a simple OLS regression of mean-differenced variables explains why this estimator is often called a within group estimator. That is, it uses only the variation within an individual's set of observations. The analysis that follows is carried out using the TSP 4.3 econometric software which provides both OLS and fixed estimates.

Table 5: Within (fixed effects) estimates (independent variable: funds gap)

Dependent variable: *net interest margin*			
Sum of squared residuals = 4014.69		R-squared = 0.8331	
Variance of residuals = 0.4319		Adjusted R-squared = 0.7915	
Std. error of regression = 0.6572			
Variance	***Estimated Coefficient***	***Standard error***	***t-statistic***
Funds gap	0.2009	0.3598	5.5834
F (2323, 9295) = 19.106, p-value = [.0000]			
Hausman test H0: RE vs. FE: CHISQ(1) = 68.995, P-value = [.0000]			

To test for the significance of the firm effects the F statistic is used. The F statistic for testing the joint significance of the firm effects is given by the formula:

$$F(n-1, nT-n-K) = \frac{\left(R_u^2 - R_p^2\right)/(n-1)}{\left(1-R_u^2\right)/(nT-n-K)}$$

where u indicates the unrestricted model and p indicates the pooled or restricted model with only a single overall constant term. Since the p-value of F obtained is sufficiently low, we reject the null hypothesis that all the coefficients are equal to zero and the evidence is strongly in favor of the firm specific effects.

The table above shows that the funds gap ratio has an econometrically and statistically significant effect on net interest margin at 1% significance level. The sign of the influence is expected. If the gap-to-assets ratio is positive, an increase in the level of interest rates will have a positive effect on net interest margin. The R-squared is high (at 0.8331).

A Hausman test for correlation between the error and the regressors can be used to check for whether the random effects model is appropriate. Under the null hypothesis of no correlation between the error and the regressors, the random effects model is applicable and its generalised least square (GLS) estimator is consistent and efficient. Under the alternative it is inconsistent. Based on a Hausman test we can conclude that of the two alternatives (fixed versus random effects), the fixed effects estimator is the appropriate choice since the p-value of chi-square obtained is sufficiently low and we reject the null hypothesis. We expected this since the data exhaust the population, and the fixed effects approach, which produces results conditional on the units in the data set, is reasonable.

We follow the analysis by adding the other variables, namely the level and variation of interest rates, the concentration measure and the asset size of each financial institution (Table 6). The model has improved significantly; the R-squared has increased from 0.8331 to 0.8616, while the sum of the squared residuals and the variance of the residuals have both been reduced. The interesting point is that the introduction of these variables has done the effects of the funds gap ratio on net interest margin statistically insignificant, although the coefficient has the right positive sign (the t-statistic is 0.1736, while the coefficient is 0.0057). It seems that there is a strong positive impact of the level of interest rates (the t-statistic is 23.2239 and the estimated coefficient 0.1451), and a negative one of the variability of interest rates (the t-statistic is -10.8739 and the estimated coefficient -0.1042) on net interest margin. For example, a 100-basis-point increase in the average level of market interest rates raises the net interest margin of banks by 0.14 percentage points. There is also a negative impact of the size of the financial institutions on net interest margin (t-statistic equals to -18.3968 while the estimated coefficient is -0.6973). Smaller banks are able to take advantage of possible increase in the yield of interest earning assets because of their relative large holdings of assets with moderately short-term maturities. Additionally, larger banks can diversify and earn profits from other activities e.g. non-interest income from fees and commissions[1]. The concentration variable is significant but negative. There is an ambiguity on the interpretation of this negative sign. If banks are able to exert market power, interest margins will be higher due to lower deposit rates, higher loan rates, or both (structure-conduct-performance hypothesis). However, the collusion effects of that power may lead to a regulation of the interest rate sensitive part of the total income with clear implications for net interest margin. In any case it is difficult to make conclusions since it is not easy to measure the level of competition.

An extension to the above models is the introduction of the lagged independent variables. Since T is quite small it would not be advisable to extend the length of the lagged values beyond the first lagged value[2]. The number of financial institutions remains constant but the time period is now 4 years, so the total number of observations is 9,296 (Table 7). The lag is presumed to reflect uncertainties, delays inherent in the decision process, and institutional rigidities.

---

[1] Sinkey (1998) shows that net interest margins vary inversely with bank asset size. The smallest banks (i.e., those not among the 1,000 largest, about 9,000) had, on average, a 173 basis point advantage over the 10 largest banks (4.41% versus 2.68%). One reason for that inverse relationship is that larger banks operate in more competitive markets, which means they must pay more for funds and cannot charge as much for loans.

[2] We can also choose the length of a lag distribution by choosing the specification with the lowest value of the Akaike Information Criterion (AIC); by implementing the OLS, the AIC in our case is quite small (-2.456).

For example, banks may adjust their sensitivity positions only a little at a time in order to determine whether their rate forecasts are materializing as planned[3]. Compared with Table 6, the R-squared is much lower (at 0.7324), while the variance of the residuals is much larger. The effect of the funds gap ratio continues to be positive (the t-statistic rises from 0.1736 when we did not include the lagged values of the independent variable to 2.9772, and the estimated coefficient from 0.0057 to 0.0823). An interesting point is that the lagged value of the funds gap has also a significant positive effect (the t-statistic is 2.1449 and the estimated coefficient 0.0552). The level of interest income is not significant (as also the lagged value of the level of interest rates), while the variability of interest rates and its lagged value are significantly negative. The t-value of the concentration measure continues to be negative and significant (-7.6168), as well as for the natural logarithm of the assets (-3.9106 and its coefficient becomes equal to -4).

Table 6: Within (fixed effects) estimates (independent variables: level and variability of interest rates, funds gap ratio, concentration and asset)

Dependent variable: *net interest margin*

Sum of squared residuals = 3330.98 R-squared = 0.8616

Variance of residuals = 0.3585 Adjusted R-squared = 0.8269

Std. Error of regression = 0.5987

Variance	***Estimated Coefficient***	***Standard error***	***t-statistic***
In. rat.	0.1451	0.0063	23.2239
Var. of int. rat.	-0.1043	0.0096	-10.8739
Funds gap	0.0057	0.0332	0.1736
Conc.	-0.0249	0.0029	-8.4654
Lnassets	-0.6973	0.0379	-18.3968

F (2323, 9291) = 17.952, p-value = [.0000]

Hausman test H0: RE vs. FE: CHISQ(5) = 181.88, p-value = [.0000]

[3] Lags may also arise in the decision process itself. "*Once interest rates begin to change, the asset-liability management committee will have to decide (forecast) whether the shift is permanent or merely an aberration. If the change is regarded as lasting over several intervals, the committee will have to develop an intermediate-term strategy which sustains the long-run strategic plan. Time may elapse between the recognision of the need to modify the intermediate-term strategy and its actual implementation*" (Graddy and Karna, 1984).

Table 7: Within (fixed effects) estimates (independent variables: level and variability of interest rates, funds gap ratio, their lagged values, concentration and asset size)

Dependent variable: *net interest margin*

Sum of squared residuals = 5203.01	R-squared = 0.7324
Variance of residuals = 0.7471	Adjusted R-squared = 0.6429
Std. Error of regression = 0.8644	

Variance	***Estimated Coefficient***	***Standard error***	***t-statistic***
In. rat.	0.0194	0.0123	1.5872
In. rat. (-1)	0.0575	0.0444	1.2946
Var. int. rat.	-0.0812	0.0047	-17.4729
Var. int. rat. (-1)	-0.2442	0.0212	-11.5201
Funds gap	0.0823	0.0277	2.9772
Funds gap (-1)	0.0552	0.0257	2.1449
Conc.	-0.1973	0.0259	-7.6168
Lnassets	-4.0079	1.0249	-3.9106

F (2323, 6964) = 5.9656 , p-value = [.0000]

Hausman test H0: RE vs. FE: CHISQ(8) = 314.96, P-value = [.0000]

6 Concluding Remarks

Monetary policy affects banks' income and profitability in many ways. Examining the effect of interest rates changes captures only one aspect of the issue. The endowment effect and the spread are captured in our analysis by the level of interest rates and the funds gap. The main findings of the paper can be summarised as follows:

- The funds gap has a significant and positive effect on net interest income; however, when we introduce the level and variability of interest rates this becomes insignificant.
- The level and variability of interest rates have significant positive and negative effects respectively on net interest income. Also, concentration and the size of each financial institution have significant negative effects.
- If we introduce lagged values the funds gap and its first lagged value is significant and positive. In this case, the variability of interest rates, the size, and the concentration measure continue to have negative sign, while the level of interest rates is no longer significant.

So, from the analysis it seems that there is a direct (level and variability) or indirect (through the balance sheet structure) influence of nominal interest rates on the net interest margin of the banks. In the long run, the direct effects have the major impact, while in the short run the indirect effects do. Rising rates increase the demand for rate-sensitive assets relative to that of rate-sensitive liabilities, while falling rates have the opposite effect. Incremental changes in bank net interest margin are directly related to the movement in market rates. However, although these effects are highly significant, there are also some other variables that have a great effect on the net interest margins, and these are the concentration of the market and the asset size of the financial institution.

In today's volatile interest-rate environment sound gap management is important. Asset decisions must not be made independent of liability decisions; they must be coordinated through a written asset-liability management policy. Since cyclical gap management involves speculation on the movement of future interest rates, the size of the gap needs to be monitored closely. With volatile interest rates, the stability of net interest margin indicates that the interest sensitive assets and liabilities are matched. If not, the monetary authorities must have estimates of the impact of market rate fluctuations on bank profitability in order to evaluate the trade-off between rate stability and the other policy goals.

References

Anderson, P. (1979). Costs and profitability of bank functions. *New England Economic Review*: 43-61.

Bourke, P. (1989). International comparisons of commercial banks profitability. (Institute of European Finance Monographs 7, University of Wales, Bangor).

Cole, R.T. (1981). Financial performance of small banks, 1977-80. *Federal Reserve Bulletin 67 (6)*: 480-485.

Davis, E., Gouzouloi, C., Spence, M. and Star, J. (1993). Measuring the performance of banks. *Business Strategy Review 4 (3)*: 1-14.

European Central Bank (2000). EU banks' income structure. (European Central Bank, April).

Flannery, M. (1981). Market interest rates and commercial bank profitability: an empirical investigation. *The Journal of Finance XXXVI (6)*: 1085-1101.

Flannery, M. (1983). Interest rates and bank profitability: additional evidence. *Journal of Money, Credit and Banking 15 (3)*: 355-362.

Graddy, D. and Karna, A. (1984). Net interest margin sensitivity among banks of different sizes. *Journal of Bank Research 14 (4)*: 283-290.

Hancock, D. (1985). Bank profitability, interest rates, and monetary policy. *Journal of Money, Credit, and Banking 17 (2)*: 189-202.

Ho, T. and Saunders, A. (1981). The determinants of bank interest margins: theory and empirical evidence. *Journal of Financial and Quantitative Analysis XVI (4)*: 581-600.

Howells, P. and Bain, K. (1998). The economics of money, banking and finance: A European text. (Addison Wesley, Longman).

Kwast, M. and Rose, J. (1982). Pricing, operating efficiency, and profitability among large commercial banks. *Journal of Banking and Finance 6*: 233-254.

Maisel, S. and Jacobson, R. (1978). Interest rate changes and commercial bank revenues and costs. *Journal of Financial and Quantitative Analysis 13*: 687-700.

Molyneux, P. and Thornton, J. (1992). Determinants of European bank profitability: A note. *Journal of Banking and Finance 16*: 1173-1178.

Opper, B. (1981). Profitability of insured commercial banks. *Federal Reserve Bulletin 67 (9)*: 657-669.

Revell, J. (1979). Inflation and financial institutions. (The Financial Times Ltd., London).

Revell, J. (1980). Cost and margins in banking: an international survey. (Organisation of Economic Co-operation and Development, Paris).

Samuelson, P. (1945). The effect of interest rate increases on the banking system. *The American Economic Review XXXV (1)*: 16-27.

Santoni, G.J. (1986). The effects of inflation on commercial banks. *Federal Reserve Bank of St. Louis 68 (3)*: 15-26.

Scott, R.H. (1966). A conditional theory of banking enterprise. *Journal of Financial and Quantitative Analysis 1*: 520-542.

Sinkey, J. (1998). Commercial bank financial management. (Prentice Hall, Fifth Edition).

Thistle, P., Mcleod, R. and Conrads, L. (1989). Interest rates and bank portfolio adjustments. *Journal of Banking and Finance 13*: 151-161.

Portfolio Management and Risk Assessment

Value at Risk Bounds for Portfolios of Non-normal Returns

E. Luciano[1] M. Marena[2]

[1] University of Turin, Department of Statistics and Applied Mathematics, Piazza Arbarello 8, 10122 Turin, Italy
[2] University of Eastern Piedmont, Department of Economics and Quantitative Methods, Via Perrone 18, 28100 Novara, Italy

Summary:
This paper studies Value at Risk (VaR) bounds for sums of stochastically dependent random variables, i.e. portfolios of correlated financial assets. The bounds hold under no restrictions on the dependence or on the marginal distributions of returns. An improvement of the bounds is given for positive (quadrant) dependent rvs. Both sets of bounds are computed for portfolios of 6 international indices. Backtesting confirms the usefulness of the approach, even with respect to other shortcuts, such as the normality assumption. For small portfolios, bounds are not over conservative.

Keywords: value at risk, dependent risks, risk management

1 Introduction

This paper studies Value at Risk (VaR) bounds for sums of stochastically dependent random variables, i.e. portfolios of correlated financial assets. The bounds are quickly computed, suitable also for fat-tailed distributions, and hold under no or very general assumptions on the dependence and marginal distributions of the portfolio components.

From the probabilistic point of view, this amounts to bounding the quantile of a sum without the independence assumption. The problem of determining the quantile of a sum exactly is not trivial, and is solved in closed form only in very specific cases, such as the perfect linear dependence or the normal one. Instead of an exact result, we exploit recent contributions on distribution functions of sums (by Denuit, Genest and Marceau (1999)) in order to provide bounds for the VaR quantile, as Embrecht, McNeil and Straumann (1999), Durrleman, Nikeghbali and Roncalli (2000) also do.

From the financial point of view, the question boils down to determining the riskiness of portfolios, whenever their assets do not have independent returns. In the financial practice, the problem is usually solved either by assuming that returns are normally distributed, or by resorting to historical or Monte Carlo simulation. As for the first approach, recent financial research has cast more and more doubts on the assumption of unconditional and joint normality: it is by now well known that financial returns usually present leptokurtic distributions, with higher peaks and fatter tails than the normal. As for the other approach, simulation is computationally intensive. Conversely, the bounds approach is very straightforward, does not need a measure of dependence and can also do without the knowledge of the marginal distributions, so that it does not suffer from model risk. It provides a quick and easy approximation on VaR, which does not assume normality, and can be improved through simulation by specifying the marginals and the exact dependency relationships.

The structure of the paper is as follows: section 2 recalls the notion of portfolio VaR and presents the analytical portfolio VaR bounds, with no assumptions on dependence. Section 3 explains how to compute them numerically. Section 4 gives an application to an internationally diversified portfolio. Section 5 introduces a class of more strict bounds, which apply in case of positive (quadrant) dependency. Section 6 uses the latter bounds on the portfolio of section 4. Section 7 evaluates the approach through backtesting. Section 8 summarizes and concludes.

2 Portfolio VaR Bounds: Analytical Results

It is well known from elementary financial theory that the return on a portfolio of n assets can be expressed as a linear combination of the returns on the n components, taking as coefficients the so-called portfolio weights. Let us denote the (random) returns with $X_1, X_2, ..., X_n$, where $X_i : \Omega \to I\!R$, $i = 1, 2, ...n$, and notice that from now on we will refer to them interchangeably as returns or random variables (rvs). If we denote the corresponding weights with θ_i, the return on the portfolio, S, can be written as $S = \sum_1^n \theta_i X_i$, where $\sum_1^n \theta_i = 1$. Let us denote with $F_S(s)$ its distribution function. For the sake of simplicity, we assume that the portfolio weights are strictly positive. For given weights, and denoting with the symbol $\stackrel{\mathrm{d}}{=}$ equality by definition, the portfolio return can be re-written as the sum of the rvs $Y_i \stackrel{\mathrm{d}}{=} \theta_i X_i$, or $S = \sum_1^n Y_i$.

It has by now become common practice to measure the riskiness of a portfolio through the VaR of its return. Given a confidence level $1 - \alpha \gg 1/2$, the VaR of the return S, $VaR_S(\alpha)$, is defined as the α-quantile of S:

$$VaR_S(\alpha) \stackrel{\mathrm{d}}{=} \inf \{s \in I\!R : F_S(s) \geq \alpha\}$$

Since S is a sum of rv, analytical results on portfolio VaR are available for some extreme cases, such as perfect positive linear correlation, perfect negative linear correlation and independency. However, these results are almost irrelevant in a financial context, since more general dependency cases usually occur.

In addition, an analytical formula for the VaR of a portfolio is available for (marginally) normally distributed returns.

Both when marginal distributions are specified to be normal and when other, more realistic assumptions (such as Student's t or mixture of normals) are introduced, provided that also the dependence structure is specified, one can resort to Monte-Carlo simulation in order to compute the VaR. Before performing the simulations, a complete knowledge (by estimation or assumption) of the parameters of each marginal distribution and of the dependency parameters[1] is needed.

As a response to the fact that Monte Carlo simulation is numerically demanding, in this section we propose bounds for VaR. The bounds can be obtained without specifying pairwise dependence measures. They can be restricted if one specifies the marginal behavior and assumes positive dependency. In this sense, our contribution can be understood as a preliminary analysis on VaR, which can be further improved by performing a full simulation study.

The bounds on VaR we use are a straightforward consequence of a result in Denuit, Genest and Marceau (1999) for the distribution function of a sum of random variables. In exploiting them we assume, for the sake of simplicity, that the rvs under exam are absolutely continuous: the results however can be appropriately modified when the continuity assumption does not hold.

By considering the bivariate case , $S = Y_1 + Y_2$, and denoting with G_1 and G_2 the distribution functions of the $Y_i, i = 1, 2$, Denuit, Genest and Marceau (1999) provide stochastic bounds for the distribution function of S. The bounds, easily obtained from a result of Makarov (1981), are:

$$F_L(s) \leq F_S(s) \leq F_U(s) \tag{1}$$

where L and U are two rvs with distribution functions $F_L(s)$ and $F_U(s)$

$$F_L(s) = \sup_{x \in IR} H(x, s)$$

$$F_U(s) = \inf_{x \in IR} \min \{G_1(x) + G_2(s - x), 1\}$$

and $H(x, s)$ is the lower Fréchet bound for the joint distribution function of Y_1 and Y_2 :

$$H(x, s) \stackrel{\mathrm{d}}{=} \max \{G_1(x) + G_2(s - x) - 1, 0\}$$

[1]Please note that dependency or association is different from linear correlation and cannot be measured, for distributions different from the normal, by the linear correlation coefficient: see Nelsen (1999).

Moreover, F_L and F_U provide the best possible bounds on F_S in the sense of stochastic dominance. It follows immediately from Denuit et alii's result that

$$VaR_U(\alpha) \leq VaR_S(\alpha) \leq VaR_L(\alpha) \tag{2}$$

for every level of confidence α.

From the point of view of financial applications, the VaR lower bound $VaR_U(\alpha)$ is particularly meaningful, since it can interpreted as the "worst possible result" at a given confidence level, when there is no knowledge of the joint distribution of returns nor of their dependence structure and only the marginal distributions are known.

For the $n > 2$ case, if one denotes with $\mathbf{1}$ the vector whose components are all equal to 1, and defines

$$T(s) \stackrel{\mathrm{d}}{=} \{\mathbf{t} \in IR^n : \mathbf{t1} = s\}$$

the stochastic bounds on F_S become

$$F_L(s) = \sup_{\mathbf{t} \in T(s)} \max \left\{ \sum_1^n G_i(t_i) - (n-1), 0 \right\} \tag{3}$$

$$F_U(s) = \inf_{\mathbf{t} \in T(s)} \min \left\{ \sum_1^n G_i(t_i), 1 \right\} \tag{4}$$

It follows that, provided one interprets L and U as the rvs with distribution functions in (3) and (4), the VaR bounds in (2) hold also for portfolios with more than 2 assets.

The usefulness of the bounds, however, depends on the possibility of computing the distribution functions of L and U, $F_L(x)$ and $F_U(x)$.

Explicit, analytical representations exist when all the Y_i belong to the same "family", such as the normal, uniform, Cauchy, exponential or (exact) shifted Pareto. For the exact Pareto family

$$F_i(x) = 1 - (x/k_i)^{-\gamma}, \qquad \gamma, k_i > 0;\ x \geq k_i \tag{5}$$

the bounds for F_S can be computed explicitly, provided that the parameter γ is the same for all the returns. In this case in fact

$$c + k_U \alpha^{-1/\gamma} \leq VaR_S(\alpha) \leq k_L \alpha^{-1/\gamma} \tag{6}$$

where $k_L \stackrel{\mathrm{d}}{=} \left(\theta_1^\beta k_1^\beta + \theta_2^\beta k_2^\beta\right)^{1/\beta}$, $\beta \stackrel{\mathrm{d}}{=} \gamma/(\gamma+1)$ and $k_U \stackrel{\mathrm{d}}{=} \max(\theta_1 k_1, \theta_2 k_2)$.

Since the same γ value for every rv Y_i is needed, Pareto analytical bounds are too specific to be meaningfully used in a financial context. Due to the restrictiveness also of the other families, it seems much more promising for financial applications to resort to numerical bounds.

3 Portfolio VaR Bounds: Numerical Results

A numerical procedure due to Williamson and Downs (1990) permits to reconstruct $F_L(s)$ and $F_U(s)$ starting from the simple knowledge of the quantiles of the $Y_i s$[2]. In the $n = 2$ case, consider $q_i(\alpha_j)$, the α_j−quantiles of Y_i, $i = 1, 2$:

$$q_i(\alpha_j) \stackrel{\mathrm{d}}{=} \inf \{v \in IR : G_i(v) \geq \alpha_j\}$$

and choose $\alpha_j = j/N$, with N arbitrarily fixed and $j = 1, 2, ...N - 1$. Let $q_i(0) = \inf \operatorname{supp} G_i$ and $q_i(1) = \sup \operatorname{supp} G_i$. When the marginal distributions have unbounded support, it is necessary to curtail the distribution so that $q_i(0)$ and $q_i(1)$ are finite. However, the choice of $q_i(0)$ and $q_i(1)$ does not affect the 0.05 and 0.01 estimated quantiles for N sufficiently large. Williamson and Downs (1990) demonstrate that the quantiles of L and U can be estimated as

$$\widehat{q}_L(\alpha_j) = \min_{j \leq k \leq N} \{q_1(\alpha_k) + q_2(1 + \alpha_j - \alpha_k)\} \tag{7}$$

$$\widehat{q}_U(\alpha_j) = \max_{0 \leq k \leq j} \{q_1(\alpha_k) + q_2(\alpha_j - \alpha_k)\} \tag{8}$$

This gives immediately the estimated distribution functions of L and U :

$$\widehat{F}_L(s) = \frac{1}{N} \sum_{j=1}^{N} I_{\widehat{q}_L(\alpha_j)}(s), \widehat{F}_U(s) = \frac{1}{N} \sum_{j=0}^{N-1} I_{\widehat{q}_U(\alpha_j)}(s)$$

where $I(s)$ is the indicator function of $[s, +\infty)$. The estimates have the property

$$\widehat{F}_L(s) \leq F_L(s) < F_S(s) < F_U(s) \leq \widehat{F}_U(s)$$

for every $q_1(0) + q_2(0) \leq s \leq q_1(1) + q_2(1)$. Also, the estimated bounds converge to the analytical bounds as $N \to \infty$ and the convergence tends to be very fast. It follows that the estimated VaR bounds are:

$$\widehat{q}_U(\alpha) \leq q_U(\alpha) \leq VaR_S(\alpha) \leq q_L(\alpha) \leq \widehat{q}_L(\alpha) \tag{9}$$

The procedure extends to the multivariate case by simple iteration.

4 An Application

This section provides VaR bounds for internationally diversified portfolios of stocks. No assumption is introduced on the pairwise dependence between stock indices, while information on the marginal quantiles, necessary in order to apply the bounds in (9), is obtained from market data.

[2]Note that the procedure can be adapted to the case when short sales are allowed i.e. $Y_i = \theta_i X_i$, $\theta_i < 0$.

	Mean	St.Dev.	Skewness	Kurtosis	J.B.
MIB30	0.09	1.60	-0.06	1.58	102.25
DAX	0.10	1.42	-0.42	2.49	48.26
UKX	0.06	1.05	-0.14	1.35	142.21
SPX	0.09	1.10	-0.41	4.01	85.71
CAC	0.10	1.34	-0.16	1.52	115.22
NKY	0.00	1.45	0.14	3.06	4.26

Figure 1: Statistics on % returns

4.1 Data

We used the daily time-series on closing values of the following indices: MIB30, DAX, UKX, SPX, CAC, NKY. The data were obtained from Bloomberg and cover the time span from December 30, 1994 to April 20, 2000: before computing the corresponding returns, we eliminated the days in which more than one market was closed, and used linear interpolation on the Japanese index in order to get data also for the days in which the corresponding value was not available, while all the others were. As a result, we obtained 1215 values for each index; for the Japanese index, around 60 data were interpolated. From these values we computed 1214 log returns. We then assumed that the returns were (historically[3]) i.i.d.: that is, we assumed the returns for different days to be observations of the same rv X_i, with distribution function $F_i(x)$.[4]

Under our simplifying assumption, Figure 1 presents some statistics for the returns on the six indices: mean, standard deviation, skewness, (excess) curtosis and the Jarque-Bera index. The skewness data are negative, as is most common, with the exception of NKY. Negative skewness and curtosis different from zero indicate departures from normality. The departure is confirmed by the values of the Jarque-Bera, which all exceed the corresponding significance level, both at the 99 and at the 99.9 level of confidence: we can therefore state, at the 99.9 % significance level, that over the time span 95-00 the six indices are not normally distributed. Again, the exception is NKY, probably because of the lack of data. The QQ-plots of the six indices show that the departure from normality depends on fat tails: as an example we report the QQ-plot of the MIB in figure 2.

[3]Following Hanksson et alii (2000), we distinguish spatial from historical dependence: the former applies to different indices, the second to values of the same index in different days.

[4]A more reliable application would have involved filtering of the data in order to take into account stochastic volatility as a departure from the (historical) iid assumption.

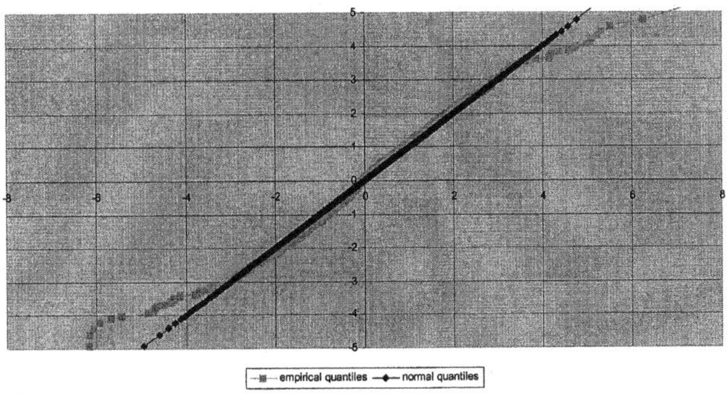

Figure 2: QQ-plot of the MIB empirical distribution versus the normal distrib- ution

4.2 VaR Bounds

4.2.1 Marginal Quantiles

For each index ($i = 1, 2, .., 6$), we computed – for $N = 1214$, $j = 1, ..., N - 1$ – the quantiles[5]of the marginal distributions, $Q_i(\alpha_j)$, defined as

$$Q_i(\alpha_j) \stackrel{d}{=} \inf\{v \in IR : F_i(v) \geq \alpha_j\}$$

in three different ways: first, we calculated simply the empirical quantiles. Secondly, we used extreme value theory (EVT) and the peak over the threshold (POT) estimation procedure in order to obtain the estimated quantiles. Thirdly, we fitted a Student's t to each index return and computed the corresponding quantiles. In all cases, we set $Q_i(0) = -30 < Q_i(1/N)$. We chose $Q_i(1)$ equal to the biggest return for the empirical and the EVT quantiles and $Q_i(1) = 30 > Q_i((N-1)/N)$ for the Student's t quantiles. Please note that these choices, consistent with Williamson and Downs (1990), do not affect the bounds at the 95 and 99% level of confidence.

As for the first approach, by putting the returns of each index in ascending order, we set $Q_i(\alpha_j)$ equal to the $j + 1$–st entry in the ordered vector of returns, for $j = 1, ..., 1214$.

As for the second approach, first of all we considered negative log returns, for the same reason as in the exact Pareto case of the previous subsection. Given that EVT results estimate quantiles over a given threshold $u(i)$, dependent in our case on the series considered, with the second approach we still used empirical quantiles below $u(i)$, while we took estimated values over

[5]Please notice that we are working on the rvs X_i, not on their "weighted" counterparts $Y_i = \theta_i X_i$: as a consequence, we do not obtain the quantiles q_i at this stage. We will transform the Q_i into q_i below.

$u(i)$. The latter were obtained from the following formula (see Embrechts, Kluppenberg and Mikocs):

$$\hat{Q}_i(\alpha_j) = u(i) + \frac{\hat{\beta}(i, u(i))}{\hat{\xi}(i)} \left(\left(\frac{n}{n_{u(i)}}(1-\alpha_j) \right)^{-\hat{\xi}(i)} - 1 \right)$$

where n is the number of data in the i-th series (1214 in our application), $n_{u(i)}$ is the number of exceedances of the threshold $u(i)$, $\widehat{\xi}(i)$ is the extreme value parameter which characterizes the EV distribution of F_i, $\widehat{\beta}(i, u(i))$ is the parameter of the generalized Pareto distribution, limit of the conditional distribution of X_i over the level $u(i)$.

The estimates of the parameters $\xi(i), \beta(i, u(i))$ for implementing it have been obtained by MLE, using Mathcad, while the threshold $u(i)$ has been chosen for each series so as to have 250 exceedances (data greater than u): $n_{u(i)} = 250$. In turn, the choice of this number of exceedances is justified by the fact that for all the series into consideration the ML estimates became stable in correspondence to that number of exceedances.

As for the third approach, due to the fat-tailed nature of the marginal returns, put into evidence by the analysis in figures (1) and (2), we assumed that marginal returns were distributed according to a Student's t:

$$F_i(s) = \int_{-\infty}^{s} \frac{\Gamma((\nu_i+1)/2)}{\sqrt{\nu_i \pi}\Gamma(\nu_i/2)} \left(1 + \frac{(u-\mu_i)^2}{\nu_i}\right)^{-(\nu_i+1)/2} du, \quad \nu_i > 0 \qquad (10)$$

where $\Gamma(\cdot)$ is the usual gamma function and $-\infty < s < +\infty$. The Student's t parameters, μ_i and ν_i, were estimated with the method of moments, by equating the theoretical and sample moments. We obtained the mean values as in figure 1, and the following ν_i values: MIB 3.29, DAX 2.98, UKX 20.82, SPX 11.63, CAC 4.52, NKY 3.82. We finally calculated the $Q_i(\alpha_j)$-quantiles by numerical inversion of (10), using Mathcad.

4.2.2 Portfolio Quantiles

We then computed the portfolio quantiles $\widehat{q}_L(\alpha_j), \widehat{q}_U(\alpha_j)$ for $\alpha_j = .05$ and $.01$, that is to say we searched for bounds on the portfolio VaR at the 95% and 99% level of confidence.

In order to do that, we needed first of all to switch from the empirical and estimated quantiles of returns, Q_i, to the corresponding quantities for weighted returns, q_i. To this end, it was sufficient to multiply the former times the weight: $q_i(\alpha_j) = \theta_i Q_i(\alpha_j)$, $\alpha_j = j/1214, j = 0, ..1214$. For our first exploration, we decided to study equally weighted portfolios of two assets: $\theta_i = \theta_z = .5$, $i = 1, 2, ...6$, $z = 1, 2, ..6$, $z \neq i$.

With the weighted quantiles $q_i(\alpha_j)$, $q_z(\alpha_j)$, using first the empirical version, then the estimated ones, we worked out the min and max search in

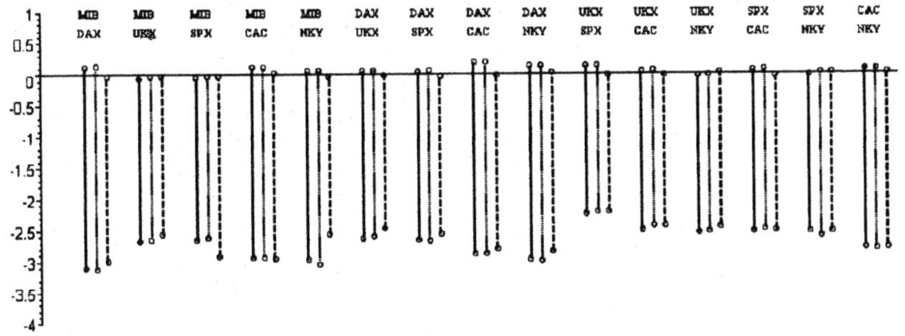

Figure 3: VaR bounds at the 95% level for equally weighted portfolios (% log returns).The first line represent the bounds obtained from the empirical quantiles while the second and third lines represent the bounds from EVT estimated quantiles and from Student's t estimated quantiles respectively

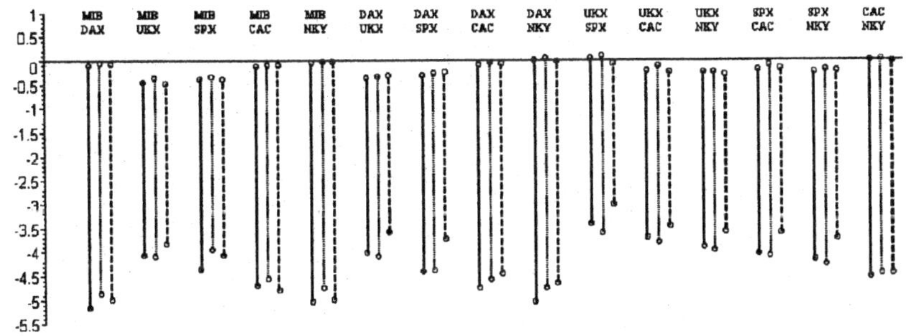

Figure 4: VaR bounds at the 95% level for equally weighted portfolios (% log returns).The first line represent the bounds obtained from the empirical quantiles while the second and third lines represent the bounds from EVT estimated quantiles and from Student's t estimated quantiles respectively

(7) and (8). By so doing, we obtained the lower and upper bounds for each portfolio and each version of the marginal quantiles. These bounds are represented, for the 95 and the 99% level of confidence, respectively in figures 3 and 4. As the reader can easily notice, the empirical bounds are not always greater (in absolute value) than the EVT or Student's ones, so that one technique cannot be said to overestimate or underestimate VaR with respect to the other. On the whole, however, empirical, EVT and Student's t bounds are fairly close to each other. In all cases, the range between portfolio VaR bounds at the 95% level was between 2.2 and 3.3 percentage points, while at the 99% level was between 2.9 and 5 points.

As mentioned in section 3.1, the VaR lower bound can interpreted as the "worst possible result" at a given confidence level, when there is no knowledge

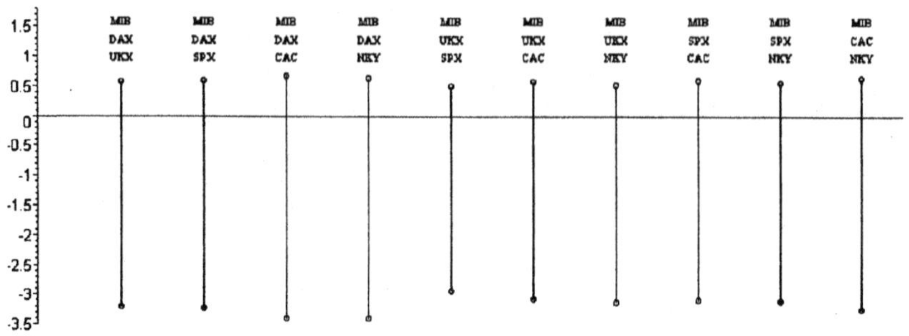

Figure 5: VaR bounds at the 95% level for equally weighted portfolios (% log returns)

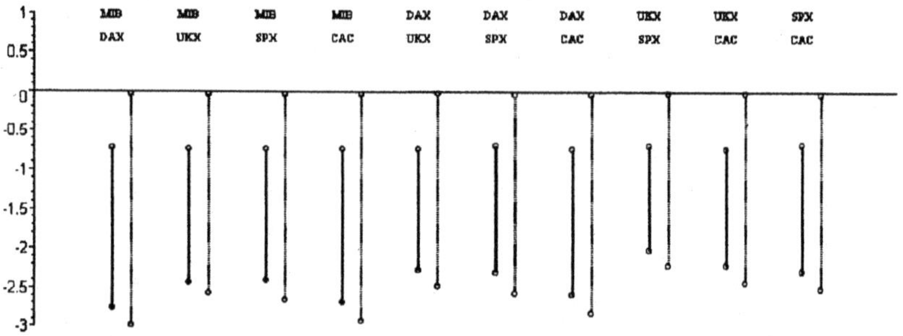

Figure 6: VaR bounds at the 99% level for equally weighted portfolios (% log returns), empirical marginal quantiles

of the joint distribution of returns nor of their dependence structure and only the marginal distributions are known.

We also extended our analysis by iteration of the technique to equally weighted portfolios of three assets: further extensions are straightforward. The method is easy to implement and fast.

Figures 5 and 6 present the 3-assets VaR bounds in case of empirical marginals. The range between the bounds is greater than in the two assets case: this is due to the fact that, excluding elliptical distributions, VaR is not subadditive. As a matter of fact, one could also verify that on our data the lower bound is smaller than the sum of the corresponding marginal quantiles, appropriately rescaled in order to consider equally weighted returns. To sum up, the extension to more than 2 assets is computationally soft, but suffers from the VaR lack of subadditivity.

5 Restricting the Bounds Through Assumptions on Association

This section provides restrictions on the bounds, for the case in which the rvs Y_i are positive quadrant dependent (PQD), i.e. have joint distribution function F greater than the independence one:

$$F(y_1, y_2) \geq G_1(y_1)G_2(y_2) \tag{11}$$

for every $(y_1, y_2) \in IR^2$.

By using (11), it is easy to prove that

$$\sup_x G_1(x)G_2(s-x) \leq F_S(s) \leq \inf_x \{G_1(x) + G_2(s-x) - G_1(x)G_2(s-x)\}$$

where both the lower and the upper bound are more strict then in section 2. We cannot state any more that the lower and upper bound are distribution functions, and consequently we cannot envisage any quantile for them. In order to provide bounds for the portfolio VaR, we therefore assume that both bounds are finite and define[6]

$$x_L^*(s) \stackrel{\mathrm{d}}{=} \arg\max_x G_1(s)G_2(s-x)$$

$$x_U^*(s) \stackrel{\mathrm{d}}{=} \arg\min_x \{G_1(x) + G_2(s-x) - G_1(x)G_2(s-x)\}$$

In addition, we denote as s_L^* the value of s such that the lower bound for the distribution function of S has value α; formally, s_L^* is the solution of the equation

$$G_1(x_L^*(s_L))G_2(s_L - x_L^*(s_L)) - \alpha = 0 \tag{14}$$

Symmetrically, we denote with s_U^* the value of s such that the upper bound for the distribution function of S has value α, i.e. the solution of

$$G_1(x_U^*(s_U)) + G_2(s_U - x_U^*(s_U)) - G_1(x_U^*(s_U))G_2(s_U - x_U^*(s_U)) - \alpha = 0$$

Under this notation, the bounds for the portfolio VaR become:

$$q_U(\alpha) \leq s_U^*(\alpha) \leq VaR_S(\alpha) \leq s_L^*(\alpha) \leq q_L(\alpha)$$

We remark that the restriction of the bounds follows from positive association, which is quite a common feature of financial returns; however, as mentioned above, it requires a full specification of the marginal distributions $G_1(x)$ and $G_2(x)$.

[6]If both marginals are absolutely continuous, $x_L^*(s)$ and $x_U^*(s)$ solve the equations

$$\frac{g_1(x_L)}{G_1(x_L)} = \frac{g_2(s-x_L)}{G_2(s-x_L)} \tag{12}$$

$$\frac{g_1(x_U)}{1-G_1(x_U)} = \frac{g_2(s-x_U)}{1-G_2(s-x_U)} \tag{13}$$

where g_1 and g_2 are the densities corresponding to G_1 and G_2.

6 The Application Revisited

This section applies the lower improvement to the portfolio VaR bounds of section 4.

In order to be able to assume PQD returns, we tested at least for positive pairwise Kendall's taus between the indices (a condition necessary but not sufficient for PQD). The Kendall's coefficients turned out to be all positive, with the exception of the Japanese index, NKY. We therefore restricted the bounds for all indices except NKY.

In addition, a full specification of the marginals was needed. We adopted here the Student fit, described in section 4 above.

If both F_i and F_z , $z \neq i$, belong to the Student's family, so that the weighted returns distributions $G_i(s) = F_i(s/\theta_i)$ are Student too, the first order condition for $x_L^*(s)$ becomes

$$\frac{\frac{1}{\theta_i}\left(1+\frac{\left(\frac{x_L}{\theta_i}-\mu_i\right)^2}{\nu_i}\right)^{-(\nu_i+1)/2}}{\frac{1}{\theta_z}\left(1+\frac{\left(\frac{s-x_L}{\theta_z}-\mu_z\right)^2}{\nu_z}\right)^{-(\nu_z+1)/2}} = \frac{\int_{-\infty}^{x_L/\theta_i}\left(1+\frac{(u-\mu_i)^2}{\nu_i}\right)^{-(\nu_i+1)/2}du}{\int_{-\infty}^{(s-x_L)/\theta_z}\left(1+\frac{(u-\mu_z)^2}{\nu_z}\right)^{-(\nu_z+1)/2}du}$$

while equation (14) is

$$\left(\int_{-\infty}^{x_L^*(s_L)/\theta_i}\left(1+\frac{(u-\mu_i)^2}{\nu_i}\right)^{-(\nu_i+1)/2}du\right)\cdot$$
$$\left(\int_{-\infty}^{[s_L-x_L^*(s_L)]/\theta_z}\left(1+\frac{(u-\mu_z)^2}{\nu_z}\right)^{-(\nu_z+1)/2}du\right)=\alpha$$

By solving numerically both equations for $\theta_i = \theta_z = 1/2$ and for each choice of i and z different from i, one gets the improved upper bound s_L^* for every equally weighted bivariate portfolio. Similarly, we solve for the improved lower bond s_U^*. The restricted bounds are reported in figures 7 and 8: as expected, since we assume positive dependence, the improvement is essentially due to a reduction in the upper bounds while the lower bounds are basically unchanged. From a practical point of view, the computational effort is not repaid by a substantial gain in accuracy for the lower bound, the "worst case" one.

7 Backtesting

This section evaluates the bounds provided in section 4 through backtesting. We computed the number of exceedences of actual losses first with respect to

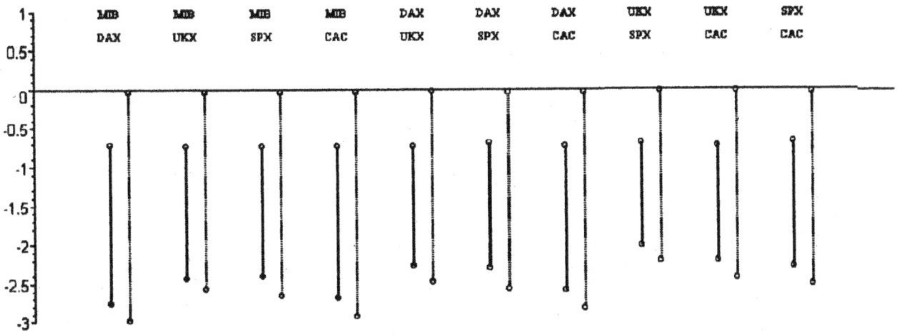

Figure 7: Restricted VaR bounds at the 95% level for equally weighted portfolios (% log returns)

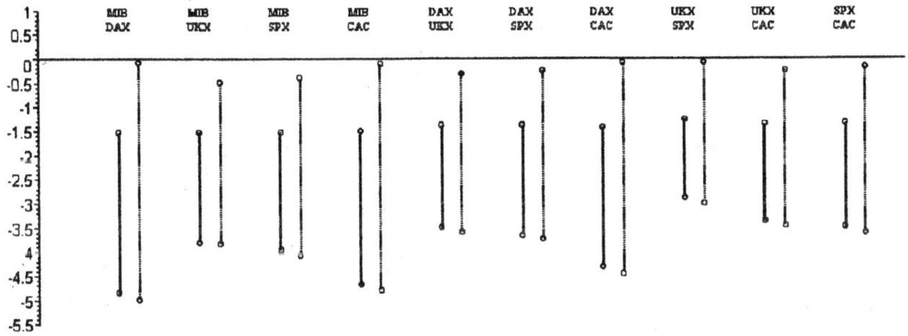

Figure 8: Restricted VaR bounds at the 99% level for equally weighted portfolios (% log returns)

	Var at 95% level		Var at 99% level	
	E Lower Bound	E Normal VaR	E Lower Bound	E Normal VaR
expected	35		7	
MIB-DAX	28 (0.11)	65 (0.00)	4 (0.12)	30 (0.00)
MIB-UKX	31 (0.24)	56 (0.00)	9 (0.23)	29 (0.00)
MIB-SPX	18 (0.00)	58 (0.00)	5 (0.22)	23 (0.00)
MIB-CAC	26 (0.06)	61 (0.00)	7 (0.49)	25 (0.00)
DAX-UKX	32 (0.29)	62 (0.00)	8 (0.36)	33 (0.00)
DAX-SPX	21 (0.01)	62 (0.00)	6 (0.35)	28 (0.00)
DAX-CAC	31 (0.24)	64 (0.00)	6 (0.35)	35 (0.00)
UKX-SPX	27 (0.08)	57 (0.00)	6 (0.35)	30 (0.00)
UKX-CAC	31 (0.24)	60 (0.00)	10 (0.13)	30 (0.00)
SPX-CAC	23 (0.02)	56 (0.00)	6 (0.35)	28 (0.00)

Figure 9: Backtesting: expected number of excedeences and number of excedeences for VaR lower bounds and for normal VaRs. p-values for a binomial test are given in brackets

the VaR lower bounds and then with respect to the VaRs under the normality assumption. We then tested whether lower bounds were too conservative, performing the binomial test in Frey and McNeil (1998).

We considered all the equally weighted portfolios of two indices, excluding NKY, since its data were partially interpolated. For each portfolio, given the time series $r_1, r_2, ..., r_{1214}$ of returns, we computed the VaR lower bound for day t, $VaR_t(\alpha)$, $t = m, ..., 1213$, with a time window of $m = 510$ days each time and $\alpha = 5\%, 1\%$. In order to obtain the lower bounds we used empirical quantiles for the marginals and the numerical procedure in section 3.2 for the portfolio. We then compared the forecast loss, $VaR_t(\alpha)$, with the actual one, r,and counted the number E of exceedences of the latter with respect to the former, i.e. $r_{t+1} < VaR_t(\alpha)$. We repeated the procedure for the corresponding VaR under the normal assumption. Figure 9 presents the number of exceedances of the lower bound and the normal approximation, for all portfolios and both levels of confidence. Our approach is closer to the expected number of exceedances, $(1213 - 510)\alpha$. As expected, E is also smaller for the lower bound than for the normal approximation.

In order to verify that the lower bound does not overestimate risk and the normal approximation does not underestimate it, we then performed a binomial test at the 95% confidence level. The null hypothesis was that each method correctly estimates VaR. If the number of violations is less than the expected number, as in most cases for the lower bound, the alternative hypothesis was that VaR is overestimated. If the number of violations was greater than expected, as with the normal and some lower bounds (namely, MIB-UKX, DAX-UKX, CAC-UKX at the 99% l.o.c.), the alternative hypothesis was that VaR is underestimated. The alternative hypotheses were confirmed by a $p-$value smaller than 0.05.

In 7 out of 10 cases at the 95% confidence level and in all cases at the 99%, the VaR lower bounds were correct, not too conservative. In all cases and at all confidence levels, the normal VaRs turned out to underestimate VaR heavily.

Backtesting then supports our approach, and in particular the interest in the lower bound, in comparison with the normal approach.

8 Summary and Conclusions

In this paper we studied Value at Risk (VaR) bounds for portfolios of correlated financial assets. The lower bound, in particular, can be interpreted as the "worst case" loss.

By exploiting recent contributions by Denuit, Genest and Marceau (1999) on distribution functions of sums, we first provided quickly-to-compute portfolio VaR bounds, as in Embrechts, McNeil and Straumann (1999) or Durrleman, V., Nikeghbali, A. and T. Roncalli, (2000); no specific assumption on the dependence structure or on the marginal distributions of returns was needed. We obtained both numerical and analytical bounds. The bounds were improved for the case of positive quadrant dependent portfolio returns and fully specified marginal distributions.

We then analyzed equally weighted portfolios of two or more international indices. In order to compute the VaR bounds, we used the daily time-series of MIB30, DAX, UKX, SPX, CAC, NKY over the time span from December 30, 1994 to April 20, 2000. The skewness, curtosis and Jarque-Bera indices and the QQ-plots provided strong evidence of departure from normality.

We found numerical bounds for portfolios of 2 and 3 assets, starting from the simple knowledge of the marginal quantiles. We evaluated the latter in three different ways: first, we calculated the empirical quantiles; secondly, we used extreme value theory (EVT) and the peak over the threshold (POT) estimation procedure; thirdly, we fitted a Student's t to each index return and computed the corresponding quantiles. All the methods considered provided fairly similar results in terms of portfolio VaR bounds. At the 95% confidence level the distance between the bounds was between 2.3 and 3.1 percentage points.

We also calculated restricted bounds, under the t assumption for marginal returns. As expected, the upper bound restriction was significant, the lower one mild: therefore the bound restriction, although interesting from a theoretical point of view, is less appealing from the practical one.

Finally we backtested our approach. Backtesting – together with a binomial test – shows not only that the 2-assets bounds are more reliable than the normal approximation, but also that they are not too conservative. The VaR lower bound appears to be moderately conservative at the 95% level and to correctly estimate VaR at the 99% level.

References

[1] Denuit, M., Genest, C.and E. Marceau, (1999) Stochastic Bounds on Sums of Dependent Risks. *Insurance: Mathematics and Economics* 25, 85-104.

[2] Durrleman, V., Nikeghbali, A. and T. Roncalli, (2000) How to get bounds for distribution convolutions? A simulation study and an application to risk management, Groupe de Recherche Opérationnelle, Crédit Lyonnais, *Working Paper.*

[3] Embrechts, P., Kluppenberg, P.and T. Mikosch, (1997). *Modeling Extremal Events for Insurance and Finance*, Springer, Berlin.

[4] Embrechts, P., McNeil, A.J. and D. Straumann, (1999), Correlation and dependency in risk management: properties and pitfalls, Department of Mathematik, ETHZ, Zurich, *Working Paper.*

[5] Frey, R., McNeil, A.J., (1998). Estimation of Tailed-Related Risk Measures for Heteroschedastic Financial Time Series: an Extreme Value Theory Approach, Department of Mathematik, ETHZ, Zurich, *Working Paper.*

[6] Makarov, G.D., (1981). Estimates for the distribution function of a sum od two random variables when the marginal distribution are fixed. *Theory of Probability and its Applications* 26, 803-806.

[7] Nelsen, R.B., (1999). *An introduction to copulas*, Springer, New York.

[8] Williamson, R.C., Downs, T. (1990). Probabilistic arithmetic I: Numerical methods for calculating convolutions and dependency bounds. *International Journal of Approximate Reasoning* 4, 89-158.

Value at Risk with High Frequency Data

E. Barucci[1] **R. Renò**[2]

[1] Dipartimento di Statistica e Matematica Applicata all'Economia, Università di Pisa
[2] Scuola Normale Superiore, Pisa

Summary:
It has been recently shown in the literature that the use of high frequency data allows very precise estimates of the daily integrated volatility for a continuous time stochastic process. A good forecasting model for daily integrated volatility is crucial for VaR estimates. Traditional models regard volatility as a latent factor; in this paper we model it as an observable quantity through an AR(n) model estimated by ordinary least squares. In spite of its simplicity, this model performs better than traditional models (GARCH(1,1) and Riskmetrics).

Keywords: Volatility, High Frequency Data, Fourier Analysis, Value at Risk

1 Introduction

The role played by volatility in most financial applications is crucial, expecially in risk management, where Value at Risk (VaR) estimates are mandatory for regulatory reasons and asset allocation decisions.

In the recent years, literature focused on the role of integrated volatility, see [1,6,7]; the importance of integrated daily volatility in VaR applications relies on two facts. First of all, it has been shown that it is possible to measure it by using intra-day data with very good precision, see [1,8,9], paralleling the use of daily returns in computing monthly volatility, see [21–23]. Second, empirical studies ([4,12]) showed that the distribution of returns divided by the square root of the integrated volatility can be well approximated by a Gaussian distribution with zero mean and variance equal to one. This fact means that VaR estimates are linked to integrated volatility forecasting, since the quantiles of the return distribution can be extracted by a Gaussian distribution with zero mean and variance given by the integrated volatility.

On the other hand, persistence properties displayed by volatility suggest that daily volatility can be forecasted with reliable precision. Typically, volatility models regard it as a latent factor which drives asset prices-returns (ARCH, GARCH models). This is the approach followed also in [10], where high frequency data are used to estimate the dynamical model for latent volatility (FIGARCH model) and to compute VaR.

In this paper, we will model directly the integrated volatility as an observable quantity through a simple $AR(n)$ model. A similar approach has been proposed by [14] and [5] in a multivariate setting. The main difference is provided by the computation method of the integrated volatility. In the above papers, integrated volatility is computed by using an equally spaced high frequency time series (typically five minute returns) as the sum of squared intraday logarithmic returns. Our procedure instead is based on the Fourier analysis methodology proposed in [17].

We will compare the performance of this model to that of the GARCH(1,1) model and to that of the Riskmetrics model proposed in [16], which is very popular among practitioners. We will show that, though the $AR(n)$ is quite simple, it performs better than traditional models in forecasting daily volatility. Our findings suggest that constructing directly a model for volatility based on the measurements of daily integrated volatility, instead of modeling volatility as a latent factor, can be a good idea.

The remainder of the paper is organized as follows: Section 2 describes the method to compute daily volatility and the three models we are going to compare; Section 3 presents the results of the comparison and Section 4 concludes.

2 Measuring daily volatility

In what follows, we will deal with univariate diffusion processes of the kind:

$$dp(t) = \sigma(t)dW(t) + \mu(t)dt, \tag{1}$$

where $W(t)$ is a Brownian motion, $\mu(t), \sigma(t)$ are allowed to be random time dependent functions. Our task is to measure and forecast the so-called integrated volatility, defined as

$$\hat{\sigma}_t^2 = \int_t^{t+1} \sigma^2(s)ds, \tag{2}$$

being one day the time unit. Recently the importance of using intra-day data in order to measure (2) has been repeatedly stressed. In [1] it is shown that using the sum of five-minute squared returns as a volatility estimate, much better results are obtained than using the simple daily squared return. This method has been employed to measure daily volatility of exchange rate time series ([2,4,19]), stock prices ([3]) and index future prices ([18]). On the same topic, see also [11,12].

In this paper, instead of the 5-minutes cumulative squared returns, we will use the integrated volatility estimator proposed in [17] which has been studied in [8,9] where it is shown that it performs better than the estimator proposed in [1]. Here we recall briefly how this volatility estimator is constructed; the interested reader is referred to [17] for details. We normalize the time window [0,T] where the time series is observed to $[0, 2\pi]$. We compute the Fourier coefficients of dp:

$$\begin{aligned} a_0(dp) &= \tfrac{1}{2\pi}\int_0^{2\pi} dp(t) \\ a_k(dp) &= \tfrac{1}{\pi}\int_0^{2\pi} \cos(kt)dp(t) \\ b_k(dp) &= \tfrac{1}{\pi}\int_0^{2\pi} \sin(kt)dp(t) \;\; k \geq 1. \end{aligned} \tag{3}$$

In [17, Theorem 1.2] it is shown that the Fourier coefficients of $\sigma^2(t)$ can be computed by means of the Fourier coefficients of dp according to

$$a_0(\sigma^2) = \lim_{n\to\infty} \frac{\pi}{n+1-n_0} \sum_{s=n_0}^{n} \frac{1}{2}\left(a_s^2(dp) + b_s^2(dp)\right) \tag{4}$$

$$a_k(\sigma^2) = \lim_{n\to\infty} \frac{2\pi}{n+1-n_0} \sum_{s=n_0}^{n} a_s(dp)a_{s+k}(dp) \tag{5}$$

$$b_k(\sigma^2) = \lim_{n\to\infty} \frac{2\pi}{n+1-n_0} \sum_{s=n_0}^{n} a_s(dp)b_{s+k}(dp), \tag{6}$$

then classical results of Fourier theory allows us to reconstruct $\sigma^2(t)$ $\forall t \in [0, 2\pi]$ by the Fourier-Féjer inversion formula:

$$\sigma^2(t) = \lim_{n\to\infty} \sum_{k=0}^{n} (1 - \frac{k}{n}) \cdot (a_k(\sigma^2)\cos(kt) + b_k(\sigma^2)\sin(kt)). \tag{7}$$

Integrating (7) between 0 and 2π we obtain our integrated volatility estimator:

$$\int_0^{2\pi} \sigma^2(t)dt = 2\pi a_0(\sigma^2). \tag{8}$$

Given a time series of N, not necessarily evenly sampled, observations $(t_i, p(t_i))$, $i = 1, \ldots N$, we will compute the integrals in (3) through *integration by parts*:

$$a_k(dp) = \frac{1}{\pi}\int_0^{2\pi} \cos(kt)dp(t) = \frac{p(2\pi) - p(0)}{\pi} - \frac{k}{\pi}\int_0^{2\pi} \sin(kt)p(t)dt. \tag{9}$$

To implement the method, we need an assumption on how data are connected. Our choice is $p(t)$ equal to $p(t_i)$ in the interval $[t_i, t_{i+1}]$ (piecewise constant). Then, the integral in (9) in the interval $[t_i, t_{i+1}]$ becomes

$$\frac{k}{\pi}\int_{t_i}^{t_{i+1}} \sin(kt)p(t)dt = p(t_i)\frac{k}{\pi}\int_{t_i}^{t_{i+1}} \sin(kt)dt = p(t_i)\frac{1}{\pi}\left(cos(kt_i) - cos(kt_{i+1})\right). \tag{10}$$

The smallest wavelength that can be evaluated is twice the smallest distance between two consecutive prices; in the case of equally spaced data, it will correspond to $k = N/2$ (Nyquist frequency). In the computation of (4) we could be tempted to stop the expansion at $N/2$. However, as shown in [8], microstructure effects, mostly due to a negative short-lived autocorrelation in the price returns (see [13,24]) prevents us to include the largest frequencies. The choice of the frequency at which to stop the expansion (4) is largely an empirical matter. The most appealing feature of this method is that it employs all the tick-by-tick observations in the integrated volatility computation. This allows to increase the precision in the estimate of (2) with respect to the sum of squared intraday returns proposed by [1], see [8]. Moreover, we completely avoid measurement biases due to an aggregating procedure of the data, see [9].

The distribution of returns divided by the square root of the daily integrated volatility can be well approximated by a Gaussian with zero mean and variance equal to one, as shown in the empirical studies in [4,12]. Than the problem of extracting the quantiles of the return distribution, i.e. the problem of calculating VaR, amounts to providing a good forecast of the integrated volatility.

Being able to measure daily volatility with good precision, we will try to model integrated volatility as an observed quantity, instead of a latent factor as in GARCH models. We will try with the simplest possible model for the time evolution of the integrated volatility, i.e. an $AR(n)$ model:

$$\hat{\sigma}_t^2 = \sigma_0^2 + \sum_{i=1}^{n} \alpha_i \hat{\sigma}_{t-i}^2 + \varepsilon_t \tag{11}$$

with $\boldsymbol{E}[\varepsilon_t] = 0$, $\boldsymbol{E}[\varepsilon_t^2] = \Sigma^2$. The parameters $\sigma_0^2, \Sigma^2, \alpha_1, \ldots, \alpha_n$ can be estimated by ordinary least squares (OLS) and $\hat{\sigma}_t^2$ is measured by the integrated volatility estimator in (8). We will use (11) to forecast future volatility.

We will compare this very simple volatility forecasting model with two models largely used by practitioners, the GARCH(1,1) model, where future volatility is estimated as:

$$\hat{\sigma}_{t+1}^2 = \psi + \alpha \cdot r_t^2 + \beta \cdot \hat{\sigma}_t^2, \tag{12}$$

where $r(t) = p(t) - p(t-1)$ is the daily return, and the model used by RiskMetrics [16], which estimates the future volatility as a sum of past realizations with exponentially declining weights:

$$\hat{\sigma}_{t+1}^2 = \frac{\sum_{i=0}^{M-1} \lambda^i \left(r_{t-i} - \frac{1}{M} \sum_{i=0}^{M-1} r_{t-i} \right)^2}{\sum_{i=0}^{M-1} \lambda^i} \tag{13}$$

where $\lambda = 0.94$.

Other authors try to model the integrated volatility, measuring it via the sum of squared intraday returns. In [14] it is found that a EMA-HAR model for the integrated volatility performs better than Riskmetrics. In [5] a tri-variate vector auto-regression (VAR), which incorporates long memory effects, is fitted on the DM-$ and Y-$ foreign exchange time-series. These authors choose a polynomial of lag 5, and find that this model performs largely better than GARCH(1,1) and Riskmetrics.

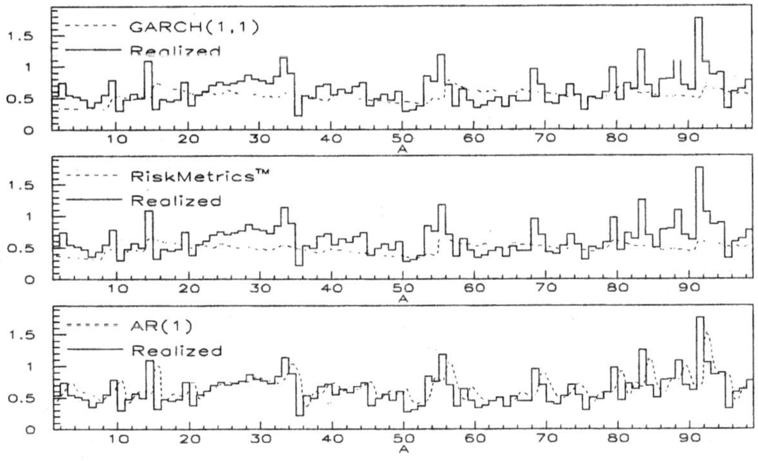

Fig. 1. Out-of-sample measured integrated volatility for the DM-$ time series (solid line) together with the forecast (dashed line) of the GARCH(1,1) model (top), Riskmetrics (center) and $AR(1)$ model (bottom).

3 Results

The data set under study in this paper is the one-year collection of bid-ask quotes of the Deutshche Mark-U.S. Dollar and Japanese Yen-U.S. Dollar exchange rates, as they appeared on the Reuters screen from October, 1st 1992 to September, 30th 1993. Each quote comes with a time stamp rounded to the nearest even second. We define a trading day to start and end 21.00 GMT; the foreign exchange market is active 24 hours per day. We excluded weekends and days with few activity due to main holidays, see [13]. This data set has been recorded and provided by Olsen & Associates, and has been extensively studied, see for example [20,15]. For any quote, we define the price to be the mid-price between the bid and the ask price. In the expansion (4) we use n as in [8], i.e. $n = 500$ for the DM-\$ time series and $n = 160$ for the Y-\$ time series. We discard days in which the observations are less than 1000 and 320 respectively; we end up with 258 daily volatility measurements for the DM-\$ time series and 259 for the Y-\$ time series. We divide our samples in 160 days for in-sample model estimate and the remainder for out-of-sample comparison.

Table 1 shows the OLS in-sample estimates of model (11), together with standard errors, the in-sample R^2 and the R^2 adjusted for degrees of freedom. For the model (12) we use the estimates given in [1], i.e. $\psi = 0.022, \alpha = 0.068, \beta = 0.898$ for the DM-\$ time series and $\psi = 0.026, \alpha = 0.104, \beta = 0.844$ for the Y-\$ time series. For the model (13) we will use $M = 160$, i.e. the largest M at our disposal.

Table 2 compares the forecasting performance. In spite of its simplicity, model (11) performs considerably better than (12) and (13). For the DM-\$ time series, it is already true with $n = 1$; however, by increasing the order of the auto-regressive model, we find better results. We interpret this finding as an evidence of long-memory effects in the volatility evolution.

For the Y-\$ time series it is necessary to employ $n = 2$, while the best result is obtained with $n = 5$. Figure 1 and 2 show the comparison between the integrated out-of-sample volatilities and the forecasts of the three models for the two exchange rate time series. Also visual inspection confirms that

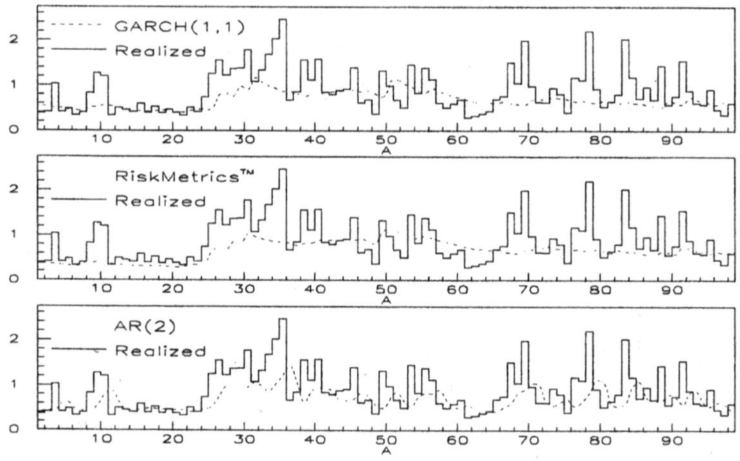

Fig. 2. Out-of-sample measured integrated volatility for the Y-\$ time series (solid line) together with the forecast (dashed line) of the GARCH(1,1) model (top), Riskmetrics (center) and $AR(2)$ model (bottom).

Table 1. OLS in-sample estimates and R^2 of model (11) for the two time series. In parenthesis, we report the standard error and the R^2 adjusted by the degrees of freedom.

Model	FX	σ_0^2	α_1	α_2	α_3	α_4	α_5	R^2 (R^2_{adj})
AR(1)	DM-$	0.209	0.728					0.245
		(0.048)	(0.051)					(0.245)
AR(2)	DM-$	0.225	0.575	0.122				0.246
		(0.047)	(0.075)	(0.072)				(0.242)
AR(3)	DM-$	0.194	0.541	-0.018	0.209			0.240
		(0.050)	(0.079)	(0.086)	(0.072)			(0.230)
AR(4)	DM-$	0.159	0.487	0.007	0.064	0.215		0.248
		(0.051)	(0.079)	(0.088)	(0.084)	(0.072)		(0.233)
AR(5)	DM-$	0.152	0.461	-0.004	0.068	0.158	0.096	0.250
		(0.053)	(0.082)	(0.088)	(0.088)	(0.084)	(0.074)	(0.231)
AR(1)	Y-$	0.299	0.381					0.150
		(0.043)	(0.074)					(0.150)
AR(2)	Y-$	0.216	0.276	0.267				0.217
		(0.048)	(0.077)	(0.077)				(0.212)
AR(3)	Y-$	0.204	0.272	0.252	0.047			0.219
		(0.051)	(0.081)	(0.080)	(0.080)			(0.209)
AR(4)	Y-$	0.192	0.264	0.250	0.032	0.052		0.221
		(0.054)	(0.081)	(0.084)	(0.083)	(0.080)		(0.206)
AR(5)	Y-$	0.188	0.262	0.248	0.030	0.046	0.023	0.222
		(0.056)	(0.082)	(0.085)	(0.087)	(0.084)	(0.081)	(0.201)

the simple $AR(1), AR(2)$ model does a good job in tracking the volatility time series. We interpret the results of our simple exercise as a confirmation that the use of high frequency data in measuring volatility, and directly modeling the integrated volatility dynamics, can substantially improve volatility forecasting, thus Value at Risk estimates.

4 Conclusions and acknowledgments

The importance of volatility measuring in risk management is increasing. Providing a good forecasting model for daily integrated volatility is essential in calculating reliable VaR estimates. Recently, it has been shown that the use of high frequency data allows to measure daily integrated volatility with high precision. This suggests that, instead of using models in which the volatility is a latent factor, like for example GARCH(1,1), one can try to model directly the dynamics of integrated volatility.

In this paper we tried to build a simple forecasting $AR(n)$ model for integrated volatility, estimated by OLS, and we showed that it performs considerably better than traditional models. We conclude that modeling directly

Table 2. Root Mean Square Error (RMSE) and Mean Absolute Error (MAE) for the $AR(n)$ model (till the best n plus one), the GARCH(1,1) model and the Riskmetrics model, for the two time series

Model	RMSE	MAE
DM-$ time series		
GARCH(1,1)	0.272	0.199
RiskMetrics	0.267	0.205
AR(1)	0.256	0.179
AR(2)	0.245	0.170
AR(3)	0.241	0.168
AR(4)	0.242	0.167
AR(5)	0.234	0.160
AR(6)	0.229	0.155
AR(7)	0.230	0.156
Y-$ time series		
GARCH(1,1)	0.508	0.346
RiskMetrics	0.486	0.344
AR(1)	0.497	0.344
AR(2)	0.472	0.340
AR(3)	0.467	0.339
AR(4)	0.461	0.335
AR(5)	0.457	0.332
AR(6)	0.458	0.334

integrated volatility, measured by high frequency data, can be a promising direction for risk management.

The authors wish to acknowledge Olsen & Associates for providing them the data set under study.

References

1. Andersen, T. and Bollerslev, T. (1998) Answering the Skeptics: Yes, Standard Volatility Models do Provide Accurate Forecasts. *International Economic Review*, 39: 885-905.
2. Andersen, T., Bollerslev, T., Diebold, F. and Labys, P. (1999) The Distribution of Realized Exchange Rate Volatility. *Journal of American Statistical Association* 96, 42-55.
3. Andersen, T. Bollerslev, T., Diebold, F. and Ebens, H. (2000) The Distribution of Stock Return Volatility. *Journal of Financial Economics*, forthcoming.
4. Andersen, T. Bollerslev, T., Diebold, F. and Labys, P. (1999), Exchange Rate Returns Standardized by Realized Volatility are (Nearly) Gaussian. *Multinational Finance Journal*, forthcoming.
5. Andersen, T. Bollerslev, T., Diebold, F. and Labys, P. (2001) Modeling and Forecasting Realized Volatility. NBER Working paper.

6. Barndorff-Nielsen, O.E. and Shephard, N. (2000) Non-Gaussian OU based models and some of their uses in financial economics. CAF Working Paper n.37, Aarhus University.
7. Barndorff-Nielsen, O.E. and Shephard, N. (2000) Econometric Analysis of Realised Volatility and it Use in Estimating Levy Based Non-Gaussian OU Type Stochastic Volatility Models. CAF Working Paper n.72, Aarhus University.
8. Barucci, E. and Renò, R. (2000) On measuring volatility and the GARCH forecasting performance. Forthcoming *Journal of International Financial Markets, Institutions and Money.*
9. Barucci, E. and Renò, R. (2000) On Measuring Volatility of diffusion processes with high frequency data. Manuscript, Università di Pisa and Scuola Normale Superiore, Pisa.
10. Beltratti, A. and Morana, C. (1999) Computing Value-at-Risk with High Frequency Data. *Journal of Empirical Finance*, 6, 431-455.
11. Bollen, B. and Inder, B. (1998), A General Volatility Framework and the Generalized Historical Volatility Estimator. Manuscript, Monash University.
12. Bollen, B. and Inder, B. (1999), Ex Post, Unconditional Estimators of Daily Volatility. Manuscript, Monash University.
13. Bollerslev, T., Domowitz, I. (1993), Trading patterns and prices in the interbank foreign exchange market. *Journal of Finance*, 48, 1421-1443
14. Corsi, F., Dacorogna, M., Muller, U., Zumbach, G. (2000) High Frequency Data Do Improve Volatility and Risk Estimation. Manuscript, Olsen & Associates.
15. Guillaime, D.M. *et al.* (1997) From the bird's eye to the microscope: A survey of new stylized facts of the intra-daily foreign exchange market. *Financc & Stochastics*, 1, 95-129.
16. J.P.Morgan, Reuters (1996) Riskmetrics - Technical Document, Fourth Edition.
17. P. Malliavin, M. Mancino, (2000) Fourier series method for Measurement of Multivariate Volatilities. Forthcoming *Finance & Stochastics.*
18. Martens, M. (2000), Measuring and forecasting stock market volatility using high-frequency data. Manuscript, University of New South Wales.
19. Martens, M. (2000), Forecasting daily exchange rate volatility using intraday returns. *Journal of International Money and Finance*, Vol.20, n.1, 1-23.
20. Muller, P. et al. (1990), Statistical Study of Foreign Exchange Rates, Empirical Evidence of a Price Change Scaling Law and Intraday Analysis. *Journal of Banking and Finance*, 14, 1189-1208.
21. Schwert, G. (1989) Why does stock market volatility change over time? *Journal of Finance*, 44: 1115-1153.
22. Schwert, G. (1990) Stock market volatility. *Financial Analyst Journal*, 46: 23-34.
23. Schwert, G. (1998) Stock market volatility: Ten years after the crash. *Brookings-Wharton Papers on Financial Services*, I, 65-114
24. Stoll, H. and Whaley, R. (1990) The Dynamics of Stock Index and Stock Index Futures Returns. *Journal of Financial and Quantitative Analysis*, 25, 441-468

Stock Index Arbitrage and Stock Price Volatility in the Athens Derivatives Exchange

M. Michalopoulos, C. Zopounidis, K. Stavroulakis

Technical University of Crete
Department of Production Engineering and Management
Financial Engineering Laboratory
Chania 73100, Greece

Summary:
This article presents the major parameters of the FTSE/ASE-20 index futures contracts which are the most actively traded in the Athens Derivatives Exchange (ADEX). In particular, data from the FTSE/ASE-20 futures and the FTSE/ASE-20 index were used in the article to examine the timing, the profitability and the risk of index arbitrage transactions. The data set includes the closing prices of the index and the settlement prices of the futures. The sample period is from April 15, 1999 to May 19, 2000. The results indicate that there were about 16 index arbitrage opportunities for all contracts. In each case, the arbitrageur had to put in practice a specific suggested strategy in order to take advantage of the opportunity. Finally, the paper discusses the postulate that stock prices have been more volatile since stock index futures contracts began trading (August 27, 1999). In fact, at monthly, weekly, daily and intra-day frequencies, volatility appears to have declined after August 1999. Furthermore, price volatility is no higher on days when future contracts are settled than on other days.

Keywords: Index futures, arbitrage, volatility, derivatives exchange

1 Introduction

The Athens Derivatives Exchange (ADEX) was established in 1998 and the futures contracts on FTSE/ASE 20 index (which are the most popular) started trading in August 1999.

Today, the following products are traded: futures on FTSE/ASE Mid 40, futures on the Hellenic Republic 10-year bond and options on FTSE/ASE 20 and FTSE/ASE Mid 40 indices.

As in all Derivatives Exchanges, there are arbitrage opportunities in ADEX as well.

In this article, the arbitrage opportunities in futures on the FTSE/ASE 20 index and the suggested strategies in order to take advantage of the opportunity are researched and analyzed.

Also, there is an attempt to analyze the statement that stock prices have been more volatile since stock index futures contracts were introduced.

The paper is organized as follows: the basic characteristics of the FTSE/ASE 20 index and futures on the FTSE/ASE 20 are presented in section 2. Section 3 provides the suggested arbitrage strategies over futures on the FTSE/ASE 20 index, and section 4 outlines the results of the application of these strategies. Section 5 examines whether the volatility of stock prices has increased since futures were introduced. Finally, section 6 contains the concluding remarks drawn from the presented analysis.

2 Stock Index Futures Contracts[1]

A stock index future gives the opportunity to someone to buy (long position) or sell (short position) the index (underlying asset) at a specific price. In that case, the buyer earns from an increase in index price, whereas an investor with a short position on the future profits from a fall in its price. The buyer or the seller can close his position by selling or buying respectively the number of contracts which make up the position. For futures on the FTSE/ASE 20 which are traded on ADEX, the underlying asset is the blue chip index FTSE/ASE 20.

2.1 The FTSE/ASE 20 index[2]

FTSE International and the Athens Stock Exchange (ASE) created the FTSE/ASE 20 index in September 1997.

2.1.1 The purpose of the index

- The FTSE/ASE 20 has been designed to provide real time measures of the Athens stock market on which index-linked derivatives can be traded.

---

[1] See Hull J. C. (2000)

[2] See Ground Rules for the Management of the FTSE/ASE 20 and the FTSE/ASE Mid 40 Indices Version 4.1 (May 2000)

- The FTSE/ASE 20 Index is the large cap index, capturing the 20 largest blue chip companies within the Athens market as agreed by the FTSE/ASE Indices Advisory Committee.

2.1.2 Independent and transparent

There are three bodies involved in the management of the index: the FTSE/ASE 20 Index Advisory Committee, FTSE International and the Athens Stock Exchange. For the independent and transparent operation of the index their decisions are based on certain ground rules. ASE is responsible for the daily operation of the index and FTSE International is responsible for auditing the index's operation and for providing advice to ASE on the treatment of complex actions.

2.1.3 Index qualification criteria

The 20 largest securities valued by adjusted market capitalization, which have passed the investibility screens comprise the FTSE/ASE 20 Index.

The shares which are eligible for inclusion in the FTSE/ASE 20 index must be listed on the Athens Stock Exchange and the company must have been established in Greece and be subject to Greek taxation. Investment companies and companies whose business is only that of holding equity are excluded. Also securities which trade on at least half of the business days in each of the preceding six calendar months and whose value of trading turnover exceeds at least 10% of the market value of the company are eligible for inclusion in the index.

2.1.4 Index algorithm and calculation method

In order to calculate the index, the following rules are used:

1. The index uses the last trade prices for securities
2. The index is calculated every minute during the opening hours of the Athens Stock Exchange using real time prices.
3. The index is displayed in two decimal points.
4. The index is calculated using the following formula:

$$\sum_{i=1}^{n} \frac{X_i W_i}{d}$$

Where:

- Xi = The latest mid price of the ith component security (or the price at the close of the index on the previous day).
- n = The number of securities in the index (=20)
- Wi = The weight for the ith component security (equal to the number of ordinary shares issued by the company)
- d = The divisor (a figure which represents the total issued share capital of the index at the base date)

2.1.5 Advantages of the index

The basic advantages of the FTSE/ASE 20 are the independence and the transparency in the index calculation and FTSE International guarantees the rules relating to changes of its constituents. Moreover, the use of the FTSE/ASE 20 index offers benefits to investors and contributes to the effective operation of the capital market. Also, the use of the index assists in fund management control and risk minimization.

2.2 Futures on the FTSE/ASE-20 index

The futures contract on the FTSE/ASE-20 index is cash settled, which means that the difference between the traded price of the contract and the closing price of the index on the expiration day of the contract are settled between the counterparties in cash. As a matter of fact, as the price of the contract changes daily, it is cash settled on a daily basis until the expiration of the contract.

The futures contract is traded in index points, while the monetary value of the contract is calculated by multiplying the futures price by the multiplier GRD 2000 (€5) per point. For example, a contract trading at 2,185 points has value of GRD 4,370,000 (€10,925).

The tick size of a future contract is 0.25 points, so that a minimum price variation is equivalent to GRD 500 (€1,25).

2.3 Use of futures on the FTSE/ASE-20 index[3]

If an investor takes position in futures on the FTSE/ASE 20 index, he has two alternatives:

1. *Hedging*. It offers the opportunity to hedge a possible position in the spot market from a falling trend of the market. Also, it can limit the loss from a

[3] See Sofianos E., (1991)

short position if there is an increase in the spot market. In both cases, hedging is easily achieved with minimum cost.

2. *Arbitrage.* Stock index futures can be used to evaluate whether an arbitrage opportunity existed. An arbitrage opportunity can exist in cases that a small movement of the FTSE/ASE 20 index (increase or decrease) takes place. Although arbitrage offers great profit opportunities, there is a risk of losing an important part of the initial capital. It must be mentioned that the risk is higher in the derivatives market than in ordinary investments (deposits, bonds and stocks). The following graph presents a comparison of the risk involved in the four investment options.

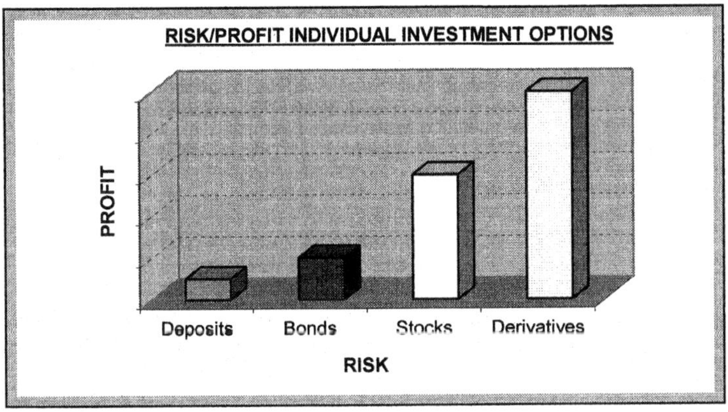

3 Stock Index Arbitrage[4]

3.1 General

An investor who takes position in the derivatives market in order to take advantage of price changes is called an arbitrageur. The difference between an arbitrageur and a hedger is that the latter's major intention is to decrease the risk of the spot market.

In general, an arbitrage opportunity appears when the relationship between the futures price and the index differs significantly. Then, an investor can buy the relatively cheap and sell the relatively expensive of the two investments.

---

[4] See Brennan M.J., and Schwartz E.S. (1990), Sofianos G., (1992) and MacKinlay A.C., and Ramaswamy K. (1988)

3.2 Estimation of the futures price

Futures price estimation is based on "the cost of carry" rule. "The cost of carry" is the difference between the spot price of a stock and its expected price at some future date and is determined by the cost of holding the stock from the present to the future date.

The formula used to calculate the theoretical price of a futures contract is:

$$F_{th} = I \cdot \left[1 + (r_b - d) \cdot \frac{T}{360} \right]$$

where:

F_{th}: Theoretical price of the future at time t

I : Index value at time t

T : Time in days from time t to expiration

r_b : Risk-free interest rate (annual deposit interest rate)
d : Dividend yield of the index portfolio

From the above equation, it is obvious that the difference between the futures price and the index depends upon the time to expiration so that, as this approaches zero, the futures price and the index converge.

However, in order to calculate the theoretical price of a future we must take into consideration the bid and ask value of the index and the market impact. As a result, the calculation of the theoretical price can be transformed into a price spread (F_{high}, F_{low}) outside of which there are arbitrage opportunities.

Therefore, using the new parameters, the theoretical prices of the future are:

$$F_{high} = \frac{I_a (1 + m_i) \cdot \left(1 + \frac{(r_a - d)T}{360} \right)}{(1 - m_f)}$$

$$F_{low} = \frac{I_b (1 - m_i) \cdot \left(1 + \frac{(r_b - r_s - d)T}{360} \right)}{(1 + m_f)}$$

where:

F_{high}, F_{low}: Maximum and minimum price of the future at time t

I_a, I_b: Bid-Ask value of the FTSE/ASE 20 index at time t

m_i: Market impact (in spot market)

m_f: Market impact (in derivatives market)

r_a: Annual lending interest rate

r_b: Annual risk free rate

r_s: Annual stock lending interest rate

d: Dividend yield of the index portfolio

T: Time in days from time t to expiration

3.3 Stock lending

Stock lending products are agreements between two parties (the lender and the borrower) through which the ownership of the stocks is transferred for a period of time and for a certain fee, from the lender to the borrower. This fee is dependent on the time that elapses until the stocks are returned to the lender and the annual interest rate. The borrower, having the ownership of the stocks, can resell them to a third party assuming the responsibility to repurchase the stocks in the future and return them to the lender. Through stock lending, the borrower can take a short position and achieve a profit or a loss, depending on the prices at which he sells and repurchases the stocks.

3.4 Basic arbitrage strategies over futures on the FTSE/ASE 20 index[5]

The following two strategies exist:

STRATEGY A: $F_{Market} > F_{High}$, which means that when the price of the contract is higher than the upper limit then:

- At time t_0 (opening position):

1. Take a short position in futures and at the same time,
2. Buy the underlying portfolio which costs X_{t0} (We assume that this amount is borrowed at r_a interest rate.)

- At time t_1 (closing position):

[5] See Brennan M.J., and Schwartz E.S. (1986), Holden G. (1990) and Habeeb G., Hill J.M. and Rzad A.J. (1991)

1. Take a long position in futures and at the same time,
2. Sell the underlying portfolio at a cost of X_{t1}.

The total profit from strategy A will be:

Profit = (Profit from closing the position in ADEX) + (Profit from selling the portfolio) – (Payment of the loan) = $(F_{t0}x2000xN - F_{t1}x2000xN) + (X_{t1} - X_{t0}) - X_{t0}\{1+\frac{t_1-t_0}{360}(r_a\text{-}d)\}$

where $N = \frac{X_{t_0}}{2000xF_{t_0}}$ number of contracts

STRATEGY B: $F_{Market}<F_{low}$, which means that when the price of the contract is lower than the lower limit then:

- At time t_0 (opening position):

1. Take a long position in futures and at the same time,
2. Sell the underlying portfolio at a cost of $X_{t0}=a_1M_{1,t0}+a_2M_{2,t0}+\ldots a_{20}M_{20,t0}$ (We assume that this portfolio is borrowed and the annual interest rate for stock lending is r_s). Then we deposit the proceeds from the selling of the stock at the annual risk free rate r_b.

- At time t_1 (closing position):

1. Take a short position in futures and at the same time,
2. Buy the underlying portfolio at a cost of $X_{t1}=a_1M_{1,t1}+a_2M_{2,t1}+\ldots a_{20}M_{20,t1}$ and return the number of stocks borrowed $(a_1,a_2,\ldots,a_{20})$.

The total profit from strategy B will be:

Profit = (Profit from closing position in ADEX)+(Profit from buying and returning the underlying portfolio) = $(F_{t1}x2000xN\text{-}F_{t0}x2000xN)+X_{t0}\{1+\frac{(t_1-t_0)}{360}(r_b\text{-}r_s\text{-}d)\}$[6] $- (a_1M_{1,t1}+a_2M_{2,t1}+\ldots a_{20}M_{20,t1})$

---

6 The amount of money that the investor will take from the deposit will be Xt0$\{1+(t_1-t_0)/360$ (rb-d)$\}$, while he will return to the stock lender Xt0$\{1+(t_1-t_0)/360$ (rs-d)$\}$. So, the available amount to buy the portfolio will be Xt0$\{1+(t_1-t_0)/360$ (rb-rs-d)$\}$.

where $N = \frac{X_{t_0}}{2000xF_{t_0}}$ number of contracts.

3.5 Portfolio selection

We choose a portfolio of shares with a previous return very close to the return of the FTSE/ASE 20 index. As a result, there is an additional risk because it is very difficult for the return of a portfolio to move in tandem with the return of the index.

Therefore, the portfolio selection was achieved with the following linear model:

$\min[z]= F-(a_1M_1+a_2M_2+a_3M_3+\ldots+a_{20}M_{20})$

where:

$$
\begin{aligned}
&0 \leq a_1 \leq 1\\
&0 \leq a_2 \leq 1\\
&0 \leq a_3 \leq 1\\
&\quad\vdots\\
&0 \leq a_{20} \leq 1\\
&a_1 + a_2 + a_3 + \ldots + a_{20} = 1
\end{aligned}
$$

F: return of FTSE/ASE-20 index

$M_1, M_2, \ldots M_{20}$: return of the shares that consist the index

$a_1, a_2, \ldots a_{20}$: the share weights

4 Application

4.1 Data description

Both futures and index prices were obtained from the Athens Derivatives Exchange. They included the settlement prices of the futures on the FTSE/ASE 20 index with delivery September 1999 to May 2000. Also the prices of the FTSE/ASE 20 index were from 15/04/1999 to 17/05/2000.

Due to the fact that no intra-day prices database existed, we used the daily settlement prices of the futures.

In order to compute the theoretical prices of the future, we used the following data:

Parameter	*Description*	*Value*
m_i	Market impact (in spot market)	1-2%
m_f	Market impact (in derivatives market)	1-2%
r_a	Annual lending interest rate	11%
r_b	Annual risk free interest rate	10,5%
r_s	Annual stock lending interest rate	2-4%
d	Dividend yield of the index portfolio	2%

In the following table the total number of arbitrage opportunities for each delivery month is presented:

Contract	*Number of opportunities*	*Date of appearance*
September 1999	1	7/9/1999
October 1999	2	7/9/1999-16/9/1999
November 1999	1	16/9/1999
December 1999	0	-
January 2000	5	16/12/1999-21/12/1999-22/12/1999-23/12/1999-24/12/1999
February 2000	2	8/12/1999-23/12/1999
March 2000	2	7/1/2000-14/1/2000
April 2000	1	17/4/2000
May 2000	2	17/4/2000-25/4/2000

The results of February 1999 and January 2000 contracts and the strategies which would have been followed are shown here below.

4.2 February 1999 contract

Date	*Market Future*	*Market Index*	*Future The. Low*	*Future The. High*	*Result*
27-Aug-1999	2807.93	2801.87	2728.74	2901.49	
30-Aug-1999	2809.70	2803.55	2728.90	2901.07	
31-Aug-1999	2780.14	2783.26	2709.15	2880.07	
1-Sept-1999	2887.73	2895.60	2818.00	2995.57	
2-Sept-1999	2925.15	2929.21	2850.19	3029.59	
3-Sept-1999	2889.95	2937.92	2858.15	3037.84	
6-Sept-1999	2950.77	3019.97	2936.39	3120.35	
7-Sept-1999	2914.55	3035.96	2951.40	3136.09	Future<TH.VA
10-Sept-1999	2968.87	3028.26	2942.33	3125.79	
13-Sept-1999	3158.45	3180.21	3088.29	3280.18	
14-Sept-1999	3162.68	3172.75	3080.49	3271.67	
15-Sept-1999	3185.94	3197.02	3103.50	3295.87	
16-Sept-1999	3218.89	3249.68	3154.05	3349.32	
17-Sept-1999	3242.83	3270.85	3174.02	3370.30	
NUMBER of OPPORTUNITIES					**1**

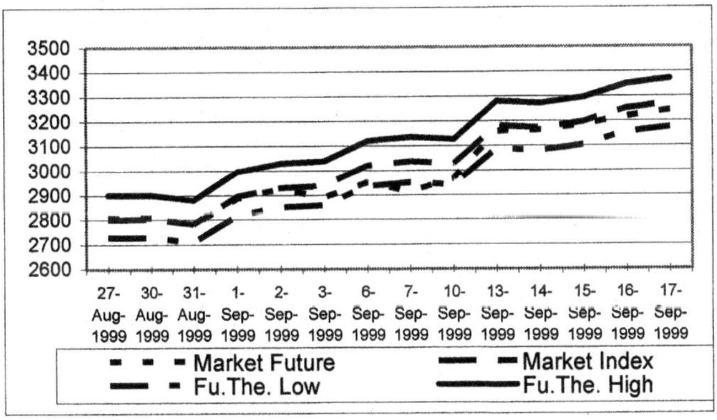

<u>Application of strategy B ($F_{Market}<F_{low}$):</u>

- On 7/9/1999 (opening position):

1. We take a long position in 10 futures, and at the same time
2. We sell the underlying portfolio that costs GRD 58,291,000 (= 2,000×10××2914.55). Therefore, according to the stock's prices on 7/9/1999, we will lend the following portfolio:

Portfolio	% Participation	Shares
NATIONAL BANK OF GREECE (CR)	22.86%	743
ALFA CREDIT BANK (CR)	14.43%	338
O.T.E SA HELLENIC TELECOMM. ORG. (CR)	7.57%	626
COMMERCIAL BANK OF GREECE (CR)	11.90%	218
VIOHALCO (CB)	5.34%	375
BANK OF PIRAEUS (CR)	6.04%	362
INTRACOM S.A (CR)	1.78%	72
PANAFON S.A (CR)	3.75%	442
EUROBANK S.A.	0.02%	1
TITAN CEMENT CO.(CR)	3.25%	108
ERGO BANK S.A. (CR)	8.68%	526
ATTICA ENTERPRISES (C)	2.57%	308
HELLENIC PETROLEUM (CR)	0.80%	118
MINOAN LINES (CR)	1.51%	102
MACED.THRACE (CR)	1.19%	69
XIOS BANK (CR)	4.78%	302
ELVAL ALUM. PROCESS (CB)	1.03%	111
ELTEX	0.56%	35
HEL.BOTTLING Co.S.A.(C)	1.65%	108
HERACLES GENERAL CEMENT Co.(CR)	0.30%	17

- Our position can be closed at any moment until the expiration. So, we must:

1. Take a short position in10 futures, and at the same time
2. Buy the underlying portfolio which costs $X_{t1}=a_1M_{1,t1}+a_2M_{2,t1}+\ldots a_{20}M_{20,t1}$ and return the number of stocks which we have lent $(a_1,a_2,\ldots,a_{20})$.

Date	Profit from Futures (GRD)	Available amount for buying the portfolio (GRD)	Portfolio cost-price (GRD)	Total Profit (GRD)
10-Sep-1999	1,086,400	58,322,574,29	58,068,377	**1,340,597**
13-Sep-1999	4,878,000	58,354,148,58	61,105,079	**2,127,070**
14-Sep-1999	4,962,600	58,364,673,35	60,971,881	**2,355,393**
15-Sep-1999	5,427,800	58,375,198,11	61,120,379	**2,682,619**
16-Sep-1999	6,086,800	58,385,722,88	62,176,307	**2,296,216**
17-Sep-1999	6,565,600	58,396,247,64	62,666,012	**2,295,836**

4.3 January 2000 contract

In the January 2000 contract, five arbitrage opportunities appeared (16/12/1999-21/12/1999-22/12/1999-23/12/1999-24/12/1999). We analyzed the first opportunity on December 16, 1999 when the price of a future in the market was higher than the upper limit of the theoretical price.

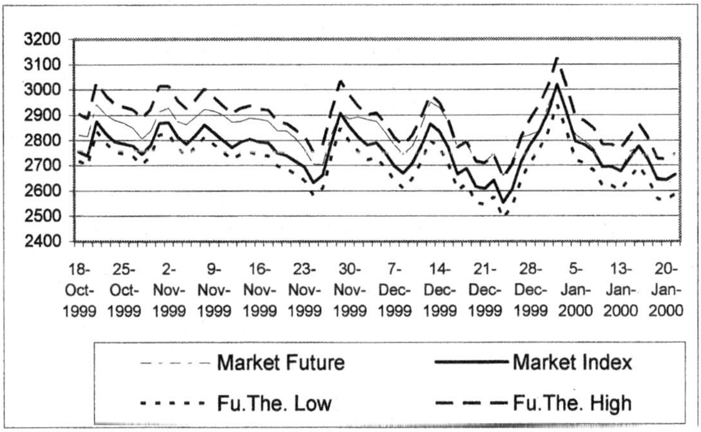

Application of strategy A ($F_{Market}>F_{low}$):

- On 16/12/1999 (opening position):

1. We take a short position in10 futures and at the same time,
2. We buy the underlying portfolio that costs GRD 55,468,800 (=10×2000××2773.44). Consequently, according to the stock's prices on 16/12/1999, we lend the following portfolio:

Portfolio	% Participation	Shares
NATIONAL BANK OF GREECE (CR)	20.94%	711
ALFA CREDIT BANK (CR)	14.23%	317
O.T.E SA HELLENIC TELECOMM. ORG. (CR)	8.57%	711
COMMERCIAL BANK OF GREECE (CR)	11.68%	290
VIOHALCO (CB)	5.47%	493
BANK OF PIRAEUS (CR)	5.99%	462
INTRACOM S.A (CR)	2.14%	91
PANAFON S.A (CR)	3.10%	445

Portfolio	% Participation	Shares
EUROBANK S.A.	0.48%	20
TITAN CEMENT CO.(CR)	3.82%	114
ERGO BANK S.A. (CR)	8.13%	474
ATTICA ENTERPRICES (C)	3.15%	338
HELLENIC PETROLEUM (CR)	1.00%	127
MINOAN LINES (CR)	2.06%	150
MACED.THRACE (CR)	1.25%	100
XIOS BANK (CR)	4.03%	323
ELVAL ALUM. PROCESS (CB)	1.81%	244
ELTEX	0.46%	22
HEL.BOTTLING Co.S.A.(C)	1.25%	95
HERACLES GENERAL CEMENT Co.(CR)	0.42%	22

- Our position can be closed at any moment until the expiration. Therefore, we must:

1. Take a long position in 10 futures and at the same time,
2. Sell the underlying portfolio costing X_{t1}.

Date	Profit from Futures (GRD)	X_{t1} (GRD)	Amount to return the loan (GRD)	*Total profit (GRD)*
17-Dec-1999	-375,200	55,848,945	55,482,667	-8,922
20-Dec-1999	1,252,600	54,400,095	55,524,269	128,426
21-Dec-1999	1,228,800	54,164,510	55,538,136	-144,826
22-Dec-1999	570,200	54,909,880	55,552,003	-71,923
23-Dec-1999	2,283,400	52,976,941	55,565,870	-305,530
24-Dec-1999	1,348,800	54,243,217	55,579,738	12,279
27-Dec-1999	-741,200	56,616,995	55,621,339	254,455
28-Dec-1999	-981,200	58,126,527	55,635,206	1,510,121
29-Dec-1999	-1,249,400	59,270,139	55,649,074	2,371,665
30-Dec-1999	-3,307,000	60,928,134	55,662,941	1,958,193
3-Jan-2000	-4,857,400	63,210,619	55,704,542	2,648,677
4-Jan-2000	-2,620,400	60,911,647	55,718,410	2,572,837
5-Jan-2000	-976,200	58,429,765	55,732,277	1,721,289
7-Jan-2000	-471,200	58,154,853	55,760,011	1,923,641
10-Jan-2000	148,000	57,606,940	55,801,613	1,953,327
11-Jan-2000	1,560,600	56,330,019	55,815,480	2,075,139
12-Jan-2000	1,549,800	56,342,096	55,829,347	2,062,549
13-Jan-2000	1,901,200	55,877,479	55,843,214	1,935,465
14-Jan-2000	385,800	56,887,635	55,857,082	1,416,354

Date	Profit from Futures (GRD)	X_{t1} (GRD)	Amount to return the loan (GRD)	*Total profit (GRD)*
17-Jan-2000	83,600	57,922,687	55,898,683	2,107,604
18-Jan-2000	1,002,200	56,856,262	55,912,550	1,945,912
19-Jan-2000	2,428,800	55,213,078	55,926,418	1,715,460
20-Jan-2000	2,528,800	55,188,582	55,940,285	1,777,097
21-Jan-2000	2,195,200	55,533,608	55,954,152	1,774,656

5 Stock Price Volatility

Various commentators have argued that trading between the stock index futures market and the stock spot market has increased the volatility of stock prices. This criticism has a long history.

For that reason the following analysis addresses two key questions: 1) Has stock price variability increased since stock index futures began trading in August 1999? 2) Are stock prices more variable on days when futures contracts are scheduled for delivery?

5.1 Percentage Changes in the FTSE/ASE 20 index: Pre- and Post-August 1999

Table 1 compares the period before and after August 1999 using monthly, weekly and daily percentage changes in the FTSE/ASE 20 Index.

Table 1: Means and Standard Deviation of FTSE/ASE 20 Index (%)

	Pre-August 1999		*Post-August 1999*		
Period	Mean	Standard Deviation	Mean	Standard Deviation	Difference in Means
Monthly[7]	2.92	7.68	-3.10	6.35	-6.02
Weekly[8]	0.97	6.14	-0.57	3.61	-0.79
Daily	0.19	2.46	-0.10	2.06	-0.29

[7] The data begin 1 March 1999 and end 18 August 2000. August 1999 is excluded.

[8] The data begin the first week of March 1999 and are percentage changes of Friday-to-Friday closing levels.

As indicated above, stock prices rose more rapidly pre-August 1999. Also, it is the variation in prices - large increases and large decreases - that appears to be the concern of the critics of derivatives.

The standard deviation is a better indicator of variation because it measures the spread of the data around the mean. Notice that the standard deviation of monthly, weekly and daily data is lower after August 1999.

Table 2 compares the variation in the index on days when futures contracts on the FTSE/ASE 20 index mature (settlement days) to the variation on all other days (non-settlement days) for the post-August 1999 period. In the case of settlement days, the data are percentage changes in the FTSE/ASE 20 index from the close on the day before a settlement day to the close on the settlement day. For non-settlement days, the data are percentage changes in the daily closing value of the index excluding the changes on settlements days.

Table 2: Means and Standard Deviation of FTSE/ASE 20 (%) index: Settlement vs. Non-settlement

	Settlement Days		***Non-settlement Days***		
Period	Mean	Standard Deviation	Mean	Standard Deviation	Difference in Means
Daily[9]	0.36	0.92	-0.12	2.10	-0.48

The Table 2 data suggest that the share prices of companies included in the FTSE/ASE 20 index were less variable on settlement (triple witching) days than on non-settlement days. These data are inconsistent with the claims of critics that stock prices are more volatile on days when futures contracts are settled.

5.2 Intra-Day Variation: Pre- and Post-August 1999

The above data measure price variation from month-to-month, week-to-week and day-to-day. Some commentators have expressed concern about intra-day movements in stock prices. The data in Table 3 examine one measure of the intra-day price variation in the FTSE/ASE 20 index for pre- and post-August 1999 data - the difference between the daily high and low of the index divided by the close and multiplied by 100.

---

[9] The data begin 1 September 1999 and end 18 August 2000.

Table 3: Means Pre- and Post-August 1999

	Pre-August 1999	*Post-August 1999*	
	Mean	Mean	Differences in Means
Intra-Day Price Spread[10]	2.89	2.47	-0.42

Table 3 indicates that intra-day volatility declined after August 1999. Table 4 examines whether the post- August 1999 intra-day price spreads have been unusually large on triple witching days. The data indicate that the mean intra-day percentage spread is lower on triple witching days than on non-settlement days.

Table 4: Means on Settlement and Non-settlement Days: Post-August 1999

	Settlement Days	*Non-settlement Days*	
	Mean	Mean	Differences in Means
Intra-Day Price Spread[Error! Bookmark not defined.]	2.17	2.49	0.32

On the whole, the data in Tables 1,2,3,4 are inconsistent with the claim that stock prices have been more volatile since stock index futures contracts began trading. In fact, at monthly, weekly, daily and intra-day frequencies, volatility appears to have declined post-August 1999. Furthermore, price volatility is no greater on days when futures contracts are settled than on other days.

6 Concluding Remarks

This article had two objectives: first to introduce a new alternative investment - the derivatives market. After the definition of the basic terms of derivatives, the arbitrage opportunities in futures on the FTSE/ASE 20 index was researched. According to the results of the research, there were a lot of opportunities during

[10] Intra-Day Price Spread $= (H - L) \cdot 100 / C$

where: H = the daily high of the FTSE/ASE 20 index
L = the daily low of the FTSE/ASE 20 index
C = the daily close of the FTSE/ASE 20 index

the first year that the futures began trading in the Athens Derivatives Exchange. In each case, the arbitrageurs had to put into practice a specific suggested strategy in order to take advantage of the opportunity.

The second objective was to the postulate that stock prices have been more volatile since stock index futures contracts began trading (August 27, 1999). In fact, price volatility appeared to have declined post-August 1999 and was no greater on days when futures contracts were settled than on other days.

References

http://www.adex.ase.gr

Brennan M.J., and Schwartz E.S., 1990, "Arbitrage in Stock Index Futures", Journal of Business, vol.63, S7-S31

Brennan M.J., and Schwartz E.S., 1986,"Optimal Arbitrage Strategies under Basis Variability", Working Paper 21-86, Graduate School of Management, UCLA

Ground Rules for the Management of the FTSE/ASE 20 and the FTSE/ASE Mid 40 Indices, Version 4.1 (May 2000)

Habeeb G., Hill J.M. and Rzad A.J., (1991) "Potential Rewards from Path-Dependent Index Arbitrage with S&P 500 Futures", *Review of Futures Markets*, Vol.10, No.1, pp.180-203

Holden G., (1990) "A Theory of Arbitrage Trading in Financial Market Equilibrium", Working Paper, University of California at Los Angeles

Hull J.C., (2000), "Options, Futures, & other Derivatives" International Edition, Fourth Edition

MacKinlay A.C., and Ramaswamy K., (1988) "Index-Futures Arbitrage and the Behavior of Stock Index Futures Prices." *Review of Financial Studies*, vol.1, pp.137-158

Sofianos G., (1991), "Comments on Potential Rewards from Path-Dependent Index Arbitrage with S&B 500 Futures", *Review of Futures Markets*, Vol.10, No.1, pp.204-206

Sofianos G., (1992) "Index Arbitrage Profitability", *Review of Futures Markets,*vol.11, pp.6-207

Evaluating Mutual Fund Performance: An Overview

K. Pendaraki, C. Zopounidis

Technical University of Crete, Dept. of Production Engineering and Management, Financial Engineering Laboratory, University Campus, 73100 Chania, Greece

Summary:
Becoming more and more popular, mutual funds have begun to play an increasingly important role in financial markets. The mutual fund performance evaluation literature is extensive but highly controversial. This paper provides a survey of mutual funds performance measures and describes the restrictions that are observed in their application. Furthermore, an overview of the literature that has been taken place in the area of mutual fund performance evaluation as well as other related works in mutual fund industry are presented.

Keywords: Mutual Funds, Performance Measures, Empirical Evidence.

1 Introduction

The mutual funds industry has experienced huge growth internationally in recent years, becoming the primary mean through which individuals and most institutions invest in capital markets. In particular, the evaluation of the performance of mutual funds has been a very interesting research topic not only for researchers, but also for managers of financial, banking and investment institutions. Although the topic of mutual funds evaluation has been ignored for many years, it recently received considerably attention from both, the financial press and the securities regulation.

According to prior research, consumers pay significant attention to the selection of the mutual funds that will best accommodate their own financial situation (Morey & Morey, 1999). Thus, it is obvious that mutual funds classes are helping investors to choose funds according to their decision policy, the risk levels that are willing to take, and their profitability needs. Today, numerous business magazines, private firms, and financial institutions are specialized in giving regular rankings and ratings of mutual funds. Furthermore, there has been a wide variety of studies regarding the development of different models and performance measures for the evaluation of mutual funds performance.

The purpose of this paper is, firstly, to present a number of mutual funds performance measures, and secondly to give an overview of the studies that have been done regarding the mutual fund performance evaluation and other related areas of the mutual fund industry. Precisely, section 2 makes a brief overview of the main mutual fund performance measures and describes the restrictions that are observed in the application of the performance measurements adjusted to risk. Section 3 outlines the empirical evidences of mutual fund performance evaluation. Finally, in section 4 the concluding remarks are discussed.

2 Brief Overview of Mutual Fund Performance Measures

The performance of mutual funds is clearly one of the most important issues on the evaluation of manager's ability to make the most appropriate investment choices and changes in the portfolio composition that will follow successfully the changes of the stock market. The performance of one mutual fund can be compared with the performance of another one, of the same or different category, giving a complete rank of mutual funds that will lead to the evaluation of the performance of investment managers.

From investors' point of view, the ranking issue is very important in order to find out the investment manager's ability. The magnitude of the return is the most important issue based on which the investor takes his decision. The evaluation of an investment can be done through the comparison of its return with the investors' opportunity cost. In the case where the investments return is lower from the opportunity cost, the investment is not successful. Another very important issue in the evaluation of mutual fund performance and the success of an investment refers to the estimation of the funds risk. It is not enough for an investor to know the magnitude of the return but it is also very important to know the risk of the return. The higher the risk of an investment the bigger the return that it gives.

This section describes several measures of mutual funds performance that have been applied in the mutual fund industry.

2.1 Return and Risk

Risk and performance measurement is an active research topic and have big interest for both investors who need to make informed decisions and to mutual fund managers whose compensation is tied to fund performance (Simons, 1998).

Investors are helped by fund managers in making more informed decisions by providing them estimates of future asset returns, risks, correlations and expected future asset allocations for their funds. The evaluation of portfolio performance is concerned with comparing the return of portfolios that have either similar or different risk levels. In either case it is very important to be more precise with the meaning of risk and return as these two basic dimensions are related with the proper evaluation of an investment.

According to the portfolio theory of Markowitz (1952), investors are taking into account both these dimensions, risk and return, in order to make an investment decision among alternative funds. According to this theory, the evaluation of mutual funds should take into account the potentialities that offer the diversification of risk among different funds and the treatment of a mutual fund as portfolio. These two dimensions of risk and return are able to well define the position of a mutual fund in a spectrum of different investments.

2.1.1 Simple measures of return

The return on a mutual fund investment in a given time period is calculated by taking into account the change in a fund's net asset value. Precisely, the change in a fund's net asset value refers to the market value of securities the funds hold, divided by the number of the fund's shares during a given time period, with the assumption of reinvestment of all income and capital-gains distributions, and dividing it by the original net asset value. The fund's monthly return is defined as follows:

$$R_{pt} = \frac{NAV_t + DIST - NAV_{t-1_t}}{NAV_{t-1}}$$

where

R_{pt} = return of mutual fund in month t

NAV_t = closing net asset value of the fund on the last trading day of the month

NAV_{t-1} = closing net asset value of the fund on the last trading day of the previous month

$DIST_t$ = income and capital distributions taken during the month.

2.1.2 Measures of risk

Investors are interested not only in fund's return but also in risks that are willing to take in order to achieve these returns. The risk concept of an investment refers to the divergence of actual and expected return. Since investors are taking risks, they receive higher levels of return with increased variability, related to risk.

Standard deviation

The variability of an investment in absolute terms is measured through the standard deviation. The standard deviation of a mutual fund is defined as follows:

$$STD = \sqrt{\frac{1}{T}\sum(R_{pt} - \overline{R}_{pt})^2}$$

where

STD = standard deviation of mutual fund in month t

R_{pt} = monthly return

$\overline{R}_{pt}$ = average monthly return

T = number of months in the period for which the standard deviation is being calculated.

The main advantage of standard deviation as a measure of variability refers to its capability of comparing alternative investments. Even though, standard deviation has some limitations, regarding the assumption of normal distribution and the comparison with only average values.

Beta coefficient

A number of performance measures are based on the capital asset pricing model, a theory that relates the expected return to a measure of risk known as beta. The beta coefficient (β) is a measure of fund risk in relation to the capital risk. It is the slope coefficient in a regression of the return of the portfolio being evaluated against the return of a proxy for the market portfolio. It is called systematic risk and the asset-pricing model implies that is crucial in determining the prices of risky assets. The beta coefficient is defined as follows:

$\beta = \text{cov}(R_{pt}, R_{Mt}) / \text{var}(R_{Mt})$

where

$\text{cov}(R_{pt}, R_{Mt})$ = covariance of monthly return of mutual fund with market portfolio

$\text{var}(R_{Mt})$ = variance of monthly return of market portfolio.

If $\beta > 1$ we have an aggressive fund which gives larger return in an increasing market but greater losses in a total decrease of the price level. If $\beta < 1$ we have a defensive fund which gives lower risk and its returns are changing more smoothly than market changes.

2.2 Risk-adjusted Performance Measures

The knowledge of the degree of risk incorporated in the portfolio of a mutual fund, gives to researchers the opportunity to use the proper performance measures of a fund. In international bibliography a variety of risk-adjusted performance measures, based in calculations of return per unit of risk or in calculations of differential returns, is developed.

2.2.1 Performance measures based on return per unit of risk

Performance measures, based on return per unit of risk, relate fund's return of a given period with the risk that incorporates. In the international bibliography, the two most commonly used measures of risk-adjusted performance are the Sharpe ratio (1966) and the Treynor ratio (1965). These two traditional total performance measures are used to measure the expected return of a fund per unit of risk.

Sharpe index

The Sharpe index or alternatively the reward-to-variability ratio is a useful measure of performance. This measure is defined as follows:

Sharpe index $= (R_p - R_f) / \sigma_p$

where

R_p = annual return of mutual fund

R_f = annual return of treasury bill (annual risk free interest rate)

σ_p = standard deviation of monthly return of mutual fund (total risk of mutual fund).

Precisely, it is defined as the difference between the fund's annual return and the pure interest rate to the standard deviation of the annual rate of return. In other words, the numerator shows the reward provided by the investor for bearing risk, while the denominator shows the amount of risk actually borne. It is obvious that this ratio is the reward per unit of variability. Furthermore, Sharpe index represents a relevant measure of mutual fund performance for investors who are not well diversified and, therefore, are concerned with their total risk exposure when evaluating mutual fund performance. The Sharpe performance measure reflects both the differences in returns to each fund and the level of mutual fund diversification.

Treynor index

This measure is defined as follows:

Treynor index $= (R_p - R_f) / \beta_p$

where

R_p = annual return of mutual fund

R_f = annual return of treasury bill (annual risk free interest rate)

β_p = systematic risk of mutual fund.

The Treynor index is obtained by simply substituting volatility (the change in the rate of return on a fund associated with 1% change in the rate of return on, say, the market portfolio) for variability in the formula of the Sharpe index. Thus, the Treynor index is similar to the Sharpe index except that performance is measured as the risk premium per unit of systematic (β_p) and not of total risk (σ_p).

The evaluation of mutual funds with those two indices show that the fund with the higher performance per unit of risk is the best managed fund, while the fund with the lower performance per unit of risk is the worst managed fund. These two indices rank portfolios but they are not able to show in percentages the overbalanced of a specific mutual fund from a passive portfolio. Furthermore, they are plagued with problems associated with the definition of risk and the need to adequately control the varying degrees of riskiness among portfolios.

2.2.2 Performance measures based on differential return

There is an alternative measure that is based on beta coefficient and gives the right answer in the question "How much higher is the return of a fund in relation with the expected one, based to its systematic risk?". This alternatively measure, which takes into explicit account the effects of risk on the returns of a portfolio, is the Jensen's performance index (1968).

Jensen alpha

The Jensen alpha measure is the intercept in a regression of the time series of fund excess returns against the time series of excess returns on the benchmark. In other words, the Jensen alpha measures a fund's average performance in excess of the return indicated by the capital asset pricing model. Both the Treynor index and the Jensen alpha assume that investors are well diversified and, therefore, they are only taking into account systematic risk when evaluating fund performance. The Jensen alpha measure is given by the regression of the following model:

Jensen model: $(R_{pt} - R_{ft}) = \alpha_p + \beta_p (R_{Mt} - R_{ft}) + \varepsilon_p$

where

R_{pt} = monthly return of mutual fund

R_{ft} = monthly return of treasury bill (monthly risk free interest rate)

R_{Mt} = monthly return of portfolio market

α_p = Jensen alpha measure

β_p = estimated risk parameter

ε_p = error term (independent normally distributed random variable with $E(\varepsilon_p) = 0$).

According to this model, inferences can be made regarding the statistical significance of the estimated performance measure. The Jensen alpha is a measure of security selection ability. Precisely, the coefficient α_p will be positive if the manager has any forecasting ability and zero if he has no forecasting ability. On the other hand, we can rule out a negative coefficient α_p by perverse forecasting ability. According to this model inferences can be made regarding the statistical significance of the estimated performance measure.

2.2.3 Modigliani measure

The three measures described above are the most common measures of risk-adjusted return but an average investor that is not well familiar with regression analysis and modern theory of finance does not easily understand the results that they are giving. For this reason, Modigliani and Modigliani (1997) proposed an alternative measure of risk adjusted performance that an average investor can easily understand. This measure is defined as follows:

Modigliani measure: $(R_p / \sigma_p) \times \sigma_I$

where

R_p = fund's average excess return

σ_p = standard deviation of fund's excess return

σ_I = standard deviation of index (S&P 500) excess return.

The fund with the highest Modigliani measure presents the highest return for any level of risk. According to this measure every portfolio is adjusted to the level of risk in its unmanaged benchmark, and then measures the performance of this risk-equivalent portfolio, comparing portfolios on the same scale. Ranking portfolios by this measure yields a score expressed in basis points. The main drawback of this measure, as well as of the Sharpe ratio, refers to its limited practical use to investors who are not in a position to use leverage in their mutual fund investing.

2.3 Fama-French Three Factor Model Alpha

Fama and French (1993) proposed a three-factor model alpha for the evaluation of fund performance that is estimated from an expanded form of the Capital Asset Pricing model (CAPM). According to this model, the security's expected return depends on the sensitivity of its return to the market return and the returns on two portfolios meant to take off additional risk factors. This model captures much of the spread in the cross-section of average returns on portfolios formed on size, book-to-market-equity, and other variables such as earnings to price, cash flow to price and long-term past return, that cause problems in CAPM model.

The Fama-French three factor model alpha is given by the following regression:

$R_{pt} - R_{ft} = \alpha_p + \beta_{p1} (R_{Mt} - R_{ft}) + \beta_{p2}\, HML_t + \beta_{p2}\, SMB_t + \varepsilon_p$

where

R_{pt} = monthly return of mutual fund

R_{ft} = monthly return of risk free interest rate

R_{Mt} = monthly return of portfolio market

HML_t = high minus low book-to-market portfolio return in month t (Fama-French book-to-market factor return)

SMB_t = small minus big size portfolio return in month t (Fama-French size factor return).

2.4 Stock Selection and Market Timing

An interesting area in finance regarding the measure of the performance of mutual funds refers to the evaluation of investment managers. This topic is of a considerable interest to investors as such information provides them useful aid for the efficient allocation of investment funds among managers. Furthermore, researchers can use information of significant evidence of superior forecasting skills if found, that would violate the efficient market hypothesis. According to Fama (1972), the forecasting skills of a fund manager are distinguished in microforecasting and macroforecasting and refer to price movements of individual stocks relative to stocks generally, called “security analysis”, and to price movements of the general stock market relative to fixed income securities, called “market timing”.

The Jensen measure does not distinguish between market timing and stock selection ability as the basis for superior performance. In the next section two traditional measures, testing the market timing and selectivity skills of an investment manager are presented. These measures that will be presented are the classic market timing model described by Treynor and Mazuy (1966) and the alternative model of market timing described by Henriksson and Merton (1981) and Henriksson (1984).

2.4.1 Treynor and Mazuy model

The Treynor and Mazuy model measures both market timing and security selection abilities. Treynor and Mazuy do so through the well known characteristic line plotting the rate of return of managed fund against the rate of return of a suitable market average. The tangent of the slope angle of the line can characterize the sensitivity of the fund examined to market fluctuations. They question that if a fund manager is able to forecast market returns, he will hold a greater proportion of the market in the case where the return on the market is high and a smaller proportion in the case where the market return is low.

According to this approach the portfolio return will be a nonlinear function of the market return. Treynor and Mazuy add a quadratic term to Jensen equation to test for market timing skills:

Treynor and Mazuy model: $(R_{pt} - R_{ft}) = \alpha_p + \beta_p (R_{Mt} - R_{ft}) + \gamma_p (R_{Mt} - R_{ft})^2 + \varepsilon_p$

where

R_{pt} = monthly return of mutual fund

R_{ft} = monthly return of treasury bill (monthly risk free interest rate)

R_{Mt} = monthly return of portfolio market

α_p = intercept term (estimated selectivity performance parameter)

β_p = estimated risk parameter

γ_p = second slope coefficient (estimated market-timing performance parameter)

ε_p = error term (independent normally distributed random variable with $E(\varepsilon_p) = 0$).

The evaluation of the performance of portfolio manager is shown through the two estimated parameters α_p and γ_p. Precisely, the parameter α_p shows the stock selection ability of the portfolio manager, the parameter β_p shows the fund's systematic risk while the parameter γ_p shows the market-timing ability of the portfolio manager. Positive values of these parameters show the forecasting ability of the portfolio manager, while negative values show the forecasting inability of the portfolio manager. Values of these parameters near to zero or zero show that the portfolio manager has no forecasting ability at all.

2.4.2 Henriksson and Merton model

Another model for the evaluation of the selectivity and timing abilities of portfolio managers refers to the Henriksson and Merton model given by the following equation:

Henriksson and Merton model: $(R_{pt} - R_{ft}) = \alpha_p + \beta_p (R_{Mt} - R_{ft}) + \gamma_p Z_{Mt} + \varepsilon_p$

where

$Z_{Mt} = \max [0, (R_{Mt} - R_{ft})]$.

Similarly to the previous model, the parameter α_p shows the stock selection ability while the parameter γ_p shows the market-timing ability of the portfolio manager. Positive values of these parameters show forecasting ability, negative values show inability of the portfolio manager while values close to zero show that the portfolio manager has no forecasting ability at all.

According to this model an investment manager attempts to forecast when the market portfolio return exceeds the risk-free rate. In the case where the forecast is for an up market, the manager adjusts the portfolio to a higher target beta while in the case where the market forecast is pessimistic, a lower target beta is used.

They present both parametric and nonparametric tests of market timing ability. The parametric tests require the assumption of either the capital asset pricing model or a multifactor return structure. In this case and based strictly on observable returns the tests permit the identification of the separate contributions from market timing ability and micro forecasting. On the other hand the nonparametric tests require knowledge of the actual forecasts or good proxy for them.

2.5 Problems in Risk-adjusted Performance Measures

The empirical literature upon the evaluation measurements of the performance of mutual fund portfolios referred to Treynor index (1965), Sharpe's index (1966), Jensen's performance index (1968), Treynor-Mazuy model (1966), Henriksson-Metron model (1981), the CAPM, and several optimization models, etc. Eventhough these performance measurements adjusted to risk have been widely used in the assessment of portfolio performance, researchers have noted several restrictions in their application (Fillipas, 1999). Some of these drawbacks are presented below:

(a) The use of a proxy variable of the theoretical market portfolio that can be criticized as inadequate. All performance measures adjusted to risk except from Sharpe ratio, are using a proxy variable that is approaching the theoretical market portfolio. The market portfolio incorporates all the assets of the examined economy. It is obvious that this portfolio is theoretical and we can have only appreciations of it. So, the use of a proxy variable can be criticized as inadequate.

(b) The evaluation of the performance of an investment manager for long and not short time periods. It must be examined if the performance of an investment manager is due to its abilities or due to lack and casual events. It is expected that in the first case the good results will continue in the future, while in the second case not. The right thing to do for the evaluation of the investment managers through the risk adjusted performance measures that are unable to distinguish the aforementioned characteristics is to be examined for a long time period.

(c) The acceptance of the assumption of borrowing and lending with the same interest rate. This assumption in relation with the previous one leads to the discrimination over a conservative portfolio and adverse an aggressive one.

(d) The validity of the CAPM. The performance measures that are using the beta coefficient (Jensen and Treynor ratio) are based in the validity of the CAPM that in many cases has been criticized. The main reason that this is happening refers to the fact that some other approach such as the Arbitrage Pricing Theory can be used.

(e) The consistency of the performance of investment managers over time, etc.

The analysis of the nature and definition of risk in the portfolio selection and management shows that the risk is multidimensional and is affected by a series of financial and stock market data, qualitative criteria and macroeconomical factors which affect the process of the capital market. Many of the models used in the past are based on unidimensional approach that does not fit to the multidimensional nature of risk (Colson & Zeleny, 1979; Hurson & Zopounidis, 1995). There are alternative approaches that provide the requisite methodology framework in handling the problem of portfolio selection and management through a realistic and an integrated methodology generated through multicriteria decision aid, neural networks, etc. A representative part of these approaches is outlined in the following section.

3 Review of Empirical Studies

The mutual fund performance evaluation literature is extensive and highly controversial. The purpose of this section is to illustrate different performance measures and approaches on the evaluation of mutual funds performance in order to give a clear view on this area.

3.1 The First Studies

Friend, et al. (1962) have done the first extensive and systematic study of mutual funds. They examined 152 mutual funds for the period 1952-58. They created an index of Standard & Poor's indexes of five securities with the elements weighted by their representation in the mutual funds sample under consideration. According to their results, there is no strong relationship between turnover rates and performance. Finally, they came up with a similar conclusion for expenses.

In 1966, there were written two papers that dominated in the area of mutual funds investment performance for the next twenty-five years. Sharpe (1966) in his study calculated the reward-to-volatility ratio for 34 funds in the period 1954-63 and found that the better performing funds tended to be those with the lower expenses. According to these results he concluded that fund's expenditure are at least partly wasted. Furthermore, he showed that performance could be evaluated with a simple theoretically meaningful measure that considers both average return and risk.

These results were very soon confirmed by the results of Jensen's research work (1968). He used the capital market line in order to calculate the alphas for his data. His sample was consisted by 115 funds in the period 1945-64. The results of his work were similar to those obtained from Sharpe's work regarding the lower performance of funds instead of the performance of a random portfolio of similar risk after the calculation of management expenses. Using the alpha measure he concluded that the examined mutual funds were on average not able to predict security prices well enough to outperform the "buy-the market-and-hold" policy. Finally, he concluded that there was little evidence that individual funds in the sample might be able to forecast prices.

3.2 Current Approaches

According to the results obtained from the traditional studies (Treyor, Sharpe, Jensen) and the general acceptance that investment managers cannot achieve returns similar to the ones of the market portfolio covering in the same time their expenses, the research interest for mutual fund was declined for almost two decades. At the end of the '80s the evaluation of the performance of investment managers earned a great deal of interest for both practitioners and academics. Many researchers used new data and methodologies and showed higher returns than the expected ones of market portfolio. Testing similar data and taking into account different parameters, others researchers could not reveal into their empirical evidence those abilities of the investments managers that characterized them as professional or not (Milonas, 1999). Some of these studies are presented below.

3.2.1 Managers beat the market

Henriksson (1984) and Chang and Lewellen (1984) in their research works found out that gross returns of mutual funds after subtracting the management expenses were similar to the expected ones of the CAPM. This fact showed that investment managers achieved higher performance from a random but well diversified portfolio insofar as the management fees to be covered. On the other hand, Ippolito (1989) according to the results obtained from his work concluded that the investment managers achieved net returns after subtracting the management expenses slightly higher from the expected ones of the CAPM.

Grinblatt and Titman (1989) in their work compared the abnormal returns of active and passive investment strategies (with and without transaction costs, fees, and expenses). The data used contained the quarterly equity holdings of a large sample of mutual funds that existed for part or all the period 1975-84. Furthermore, a benchmark was used in order to conciliate the possibility that well-known passive strategies could drive their result. According to this benchmark and applying the Jensen measures they found out that superior performance might exist among aggressive-growth and growth funds and those funds with the smallest net asset values. These funds have the highest expenses so their actual returns did not show abnormal performance. A general conclusion of their work indicates that the risk-adjusted gross returns of some funds were significantly positive. Grinblatt and Titman assert that their approach gives better evaluations of mutual fund managers' performance.

Gallo and Swanson (1996) evaluated the performance of 37 U.S. international equity mutual funds in the period 1985-93 through Sharpe and Jensen performance measures. Precisely, the Jensen measure was computed using an international two-index model and a two-factor International Arbitrage Pricing Theory (IAPT)-based performance methodology. According to the results from the index model, the international funds perform as well as the market proxy. On the other hand, the IAPT model showed superior investment performance by the international funds. The two models produce different performance rankings. Finally, the IAPT model explained better the international equity return-generating process in the examined period, illustrating statistical significant mean alphas.

3.2.2 Managers cannot beat the market

Elton et al. (1993) studied the informational efficiency of mutual fund performance using the methodology and the data set of Ippolito's work (1989) for the period 1965-84. Ippolito's work is the first test of expanded definition of market efficiency using data on managed portfolios. They applied the standard index performance measurement model and two new indexes incorporated into the standard model and found that returns on S&P stocks, returns on non-S&P stocks and returns on bonds are significant factors in performance assessment. In the case were they corrected for the impact of non-S&P assets on mutual funds returns they concluded that mutual funds do not earn returns that validate their information acquisition costs. Finally, they showed that mutual fund managers underperform passive portfolios, while the returns of mutual funds could not exceed the expected ones.

Elton et al. (1996) examined the predictability for stock mutual funds using risk-adjusted returns. They used a sample free of survivorship bias of 188 mutual funds covering the period 1977-1993. Precisely, the fund's risk-adjusted performance was based on alpha measure from a four-index model. They found out that past performance was predictive of future risk-adjusted performance. Furthermore, they formed a combination of actively managed portfolios with the same risk as a portfolio of index funds but with higher mean return. Finally, the portfolios selected had small risk adjusted returns during the examined period where mutual funds in general had negative risk-adjusted returns.

Mutual fund's returns that exceed the expected ones according the standard asset pricing models came up from simple trading rules and investment techniques, which are also taking advantage the time-series anomalies on prices. Grinblatt et al. (1995), in their study examined momentum investment strategies on quarterly holdings of 155 mutual funds for the period 1975-84. They concluded that the simple momentum strategies of buying past winners and selling past losers would generate abnormal performance with either of the Grinblatt and Titmant (1989, 1993) performance measures, as well as with any of the more traditional measures.

One of the reasons that fund managers achieve abnormal performance is due to the fact that the examined sample period covers the 1970-decade that is characterized by successful investment strategies. Malkiel (1995) examined both the 70's and 80's decades and found out that the abnormal performance of investment managers is exclusively due to the 1970 decade. So a general conclusion refers to the fact that the abnormal performance of investment managers is related to the time period that in each case is examined.

Another reason and the most important one refers to the way the examined sample is formed. Most of the researchers include in their sample those mutual funds that operate the whole examined period. The rest mutual funds are excluded from the analysis because they had not been in existence for the whole period examined due to unsuccessful performance and bad investment policy. This technique followed by researchers lead to overestimation of the examined mutual fund performance since the funds that are incorporating to their research are by definition the most successful or the less unsuccessful ones.

3.3 Performance Persistence

One very interesting and important area in measuring mutual fund performance refers to the investigation of persistence in funds' performance. There is a series of empirical studies in support to the efficient markets hypothesis that past performance is no guide to future performance. Even though, a series of empirical studies reveal that the relative performance of equity mutual funds persists from period to period. Several research works examined the predictability of mutual funds as part of mutual fund performance while others directly examined the persistence in funds performance.

As far as the traditional works concern, Sharpe (1966) showed that the Sharpe index and particularly the Treynor index demonstrate the predictability of performance and the persistence of relative performance from period to period. On the other hand, Jensen (1968, 1969) did not recognize persistence of significantly good performance in the followed period.

3.3.1 Evidence of performance persistence

Hendricks et al. (1993) studied funds performance through quarterly return data in the period 1974-88. They found out that the persistence of relative superior fund performance proves to be significant, although it refers to a short period of time. Funds delivering sustained short-run superior performance have "hot hands". Portfolios of recent poor performers did significantly worse than standard benchmarks while those of recent top performers did better. Furthermore, the difference in risk-adjusted performance between the top and the bottom octile portfolios is six to eight percent per year.

Goetzmann and Ibbotson (1994) examined whether the best-performing managers in the past are the best-performing managers in the future too. They tested the repeat-winner pattern over successive one and two year intervals from 1976 to 1988. According to the results of their work, it appeared that past alphas predict future alphas, implying either that managers reveal superior relative performance, or at least that some managers appears to be superior at controlling this common performance measure.

Brown and Goetzmann (1995) studied mutual funds performance persistence using absolute and relative benchmarks for the years 1976 through 1988, for equity mutual funds. Their data set was free of survivorship-bias and indicates that relative risk-adjusted performance of mutual funds persists but mostly due to funds that lag the S&P 500. Furthermore, they applied a probit analysis and found out that poor performance increases the probability of disappearance. They decomposed the persistence effect year-by-year and showed that the performance pattern depends on the examined period and it is also correlated across managers and their common strategies.

Grinblatt and Titman (1992), Elton et al. (1993) and Elton et al. (1996) in their studies ensured higher run funds performance persistence than the three previous studies which are referred to short run performance persistence. They examined the performance persistence of mutual funds for five to ten years period of time attaching the ability of investment managers to the fact that they can be properly informed and make good choices.

Suggestively, it is presented the work of Grinblatt and Titman (1992), which examined the presence of mutual fund performance persistence for 279 funds that existed from 1974 to 1984. The performance measure used in their study is an extension of the Jensen measure (1968) and is computed relative to an eight-portfolio benchmark. This multiple portfolio benchmark was formed on the basis of securities characteristics. The results of their work showed that past performance of a fund provides useful information for investors that are planning to invest in the future. Furthermore, they pointed out that the performance persistence is consistent with the ability of fund managers to earn abnormal returns.

3.3.2 Slight evidence of performance persistence

Malkiel (1995) examined the returns from investing in equity mutual funds for the period 1971-91 and was one of the first researchers that contested to the existence of mutual funds performance persistence. He followed the methodology of Goetzmann & Ibbotson (1994) and spoke about "winner" mutual funds that achieved in a specific year returns higher than the median returns and "loser" mutual funds that achieved returns lower than the median ones.

Carhart (1997), used a sample of equity mutual fund for the period 1962-93 free of survivor bias in order to demonstrate that common factors in stock returns and investments expenses almost completely explain persistence in fund's mean and risk-adjusted returns. He used the CAPM described in Sharpe (1964) and Litner (1965), and his four-factor model (Carhart, 1995) constructed by the three-factor model of Fama and French (1993). This four-factor model is consistent with a model of market equilibrium with four risk factors (systematic risk, market capitalization, book-to-market equity, and one-year return momentum) and explains considerable variation in returns. The results of his work do not support the existence of skilled mutual fund managers.

Kahn and Rudd (1995), investigated the performance persistence of 300 equity mutual funds for the period 1983-93 and 195 fixed income funds for the period 1986-93. They used style analysis in order to separate fund total returns into style and selection components. For each fund type they examined the performance persistence from one period to the next using regression analysis and contingency tables. The results indicated that there is no evidence for persistence of performance among equity mutual funds, but on the other hand, they found evidence of persistence of fixed income selection returns and information ratios even after accounting for expenses.

3.4 Market Timing and Selectivity

Several studies have examined the market-timing and stock selection abilities of mutual fund managers. The academic literature on this area among others concerns the existence or lack of superior performance of investment managers or even the correlation between selectivity and timing abilities. A review of these approaches is presented below.

Henriksson (1984) examined the forecasting ability of mutual fund managers using both the parametric and nonparametric techniques presented by Henriksson and Merton (1981). Precisely, he examined the market-timing ability of 116 open-end mutual funds for the period 1968-80. The results obtained did not supported the hypothesis that mutual fund managers are able to follow an investment strategy that successfully times the return on the market portfolio. In addition, he found that only three funds had significantly positive estimates of market-timing ability. Herniksson found evidence of dynamic heteroscedasticity, but correcting it did not alter his conclusion.

Chang and Lewellen (1984) in their research work used the Henriksson and Merton (1981) statistical procedure to test jointly for the presence of either superior market timing or security selection ability in managed portfolios in a sample of 67 mutual funds covering the period 1971-1979. According to their results only few fund managers appear to have demonstrated much market-timing and security selection skills. The general conclusion in which they came up refers to the inability of mutual funds to outperform a passive investment strategy, a conclusion that was consistent with prior literature.

Lehmann and Modest (1987) in their research work examined the returns of 130 mutual funds over the period 1968 through 1982. They combined the APT performance evaluation method with the model of Treynor and Mazuy and tried to ascertain whether conventional measures of abnormal mutual fund performance are sensitive to the benchmark chosen to measure normal performance. Furthermore, they employed the standard CAPM benchmarks and a variety of APT benchmarks in order to give an answer to the previous question. They found statistically measured abnormal timing and selectivity performance by mutual funds. Another interesting finding of their work referred to the fact that performance measures are quite sensitive to the benchmark chosen and a large number of negative selectivity measures.

Cumby and Clen (1990) examined the performance of 50 U.S-based internationally diversified mutual funds for the period 1982 to 1988. They used two performance measures, the Jensen measure and the positive weighting measure, proposed by Gninblatt and Titman (1989), and found that there is no evidence that funds provide investors with performance that surpasses that of a broad, international equity index over the examined period. Among these measures they searched for market timing ability with the Capital international World Index as the benchmark through the Treynor and Mazuy (1966) model. They found out thát there exists perverse market timing by fund managers.

Koh, et al. (1993), examined the market timing abilities of fund managers using both parametric and nonparametric tests of Henrissson and Merton (1981) on a sample of fund managers in Singapore for the period 1980-1987. Using the nonparametric test, there were found market timing abilities among the managers, while the use of parametric test led to the opposite conclusion. They tried to reconciliate the empirical results, but eventually they found that they could not rule out the existence of market timing abilities of fund managers in Singapore the examined period.

Bello and Janjigian (1997) in their research work used an extended version of the Treynor-Mazuy model in order to examine the market-timing and stock selection ability of equity mutual funds. The methodology used, was designed to correct the leverage characteristics of mutual fund holdings, and holdings of bonds, controlling for the inclusion of non-S&P 500 assets in mutual fund portfolios. They found positive and significant market-timing abilities for 633 mutual funds in 1984-94 period. Furthermore, they found that security-selection abilities are significantly positive and cross-sectional correlations between market timing and selectivity are significantly negative. Finally, they found out that monotonic correlation between turnover and market timing is positive, and negative between turnover and selectivity. None of these findings were evident with the original Treynor-Mazuy model, while the extended model used was correctly specified and unbiased.

3.5 Alternative Approaches

Most of the studies concerning the mutual fund performance evaluation utilise linear models as the primary tools and it is possible that they do not capture the complexity presented in the data. At this part of our study it is presented the problem of mutual fund performance evaluation by utilising alternative approaches, most of them free from the aforementioned criticism.

Brockett et al. (1992) in their empirical analyses of mutual fund investment strategies used 383 open-end mutual funds for which they obtain ratings for 1984 through 1988. They took into account the publicly announced strategies of individual funds in order to form ex ante risk classifications that are employed in examining ex post performance. They tested a chance constrained programming approach in order to maximise the possibility of the performance of a mutual fund portfolio to exceed the performance of the S&P 500 index formalising risk and return relations. Negative relations between risk and return held in every year of the examined period.

Chiang et al. (1996) examined 101 U.S. mutual funds for the period 1981 through 1986. They used an artificial neural network method in order to develop forecasting models for the prediction of end-of-year net asset values of mutual funds, taking into account historical economic information. Precisely, various economic variables like gross national product, consumer price index and inflation rate were used as exogenous variables for the models. They compared their forecasting results to those of traditional econometric techniques and concluded that neural networks significantly outperform regression models in situations with limited data availability.

Murthi et al. (1997) examined the efficiency of mutual fund industry by different investment objectives using a data set of 731 mutual funds for the third quarter of 1993. They tried to overcome the limitations of traditional indices, proposing a new measure of performance that is calculated through the data envelopment analysis (DEA), a well-known method in operation research. Precisely, they proposed a DEA portfolio efficiency index that compares the index of each portfolio to the best portfolio available in the mutual fund's class establishing a benefit/cost non-parametric analysis. Given the transaction cost structure and risk taken, they evaluated which fund generates the highest return in the same class. Using this approach they found strong evidence that the mutual funds are all approximately mean-variance efficienct.

Indro et al. (1999) used into their research work artificial neural networks in order to predict mutual fund performance. They utilised a total of 559 equity mutual funds of three different investment styles over a three-year time period, 1993-95. They used a multi-layer model and a non-linear optimiser taking into account funds specific historical operating characteristics in order to forecast mutual funds' risk adjusted return. They concluded that whether the neural network approach is superior to linear models for predicting mutual fund performance depends on the style of the fund.

Morrey and Morrey (1999) into their mutual fund performance appraisals examined 26 mutual funds over a three, five, and ten-year time horizons (sample period ended on 1995). They used two basic quadratic programming approaches in order to identify those funds that are strictly dominated, regardless of the weightings on the different time horizons examined and relative to their mean returns and risks. These two quadratic programming approaches can be characterized as efficient frontier approaches, motivated by data envelopment analysis. Finally, in contrast to other mutual funds studies these approaches endogenously determine a custom tailored benchmark portfolio to which each mutual fund's performance is compared.

Pendaraki and Zopounidis (2000) into their research work tried to solve the ranking problem of the performance of mutual funds through Promethee II (Preference Ranking Organisation Method) method that is originated from the field of the multicriteria decision aid. A sample of Greek domestic equity mutual funds derived from the Association of Greek Institutional Investors for the period 1997-1999 is used to rank funds according to their performance. This method was applied in order to rank mutual funds from the best to the worst ones based on three generalized criteria, the combination of them, and nine criteria of mutual fund performance through several different scenarios. The results indicate that there are no significant differences in the rankings of the mutual funds for the three years examined.

3.6 The Most Recent Studies

At this part of our study it is presented a sample of the most recent works on the topic of mutual fund performance as well as other related works that have been done on this area.

Dalhquist et al. (2000) studied the relation between fund performance and fund attributes in the Swedish market. They examined 130 equity mutual funds for the period 1993-97. According to their work, performance is measured as the alpha in a linear regression of fund returns on several benchmark assets, allowing for time-varying betas. They came up with the conclusion that good performance occurs among small equity funds, low fee funds, funds whose trading activity is high and in few cases funds with good past performance.

Wermers (2000) in his study performed a comprehensive analysis of mutual fund industry through a new database that allows an analysis of mutual funds in both the stock holdings level and the net return level from 1975 to 1994. He decomposed performance into several components to analyze the value of active fund managers. According to the results of the application of the performance decomposition methodology (characteristic selectivity and timing measures, average style measure, and execution costs) followed in this study, funds that hold stocks outperform the market, whereas their net returns underperform the market. Thus, funds include stocks to cover their costs. Finally, there is evidence that supports the value of active mutual fund management.

Pilpot et al. (2000) studied the performance persistence and management skill in 73 non-conventional bond mutual funds over the period 1988-97. They calculated the funds Sharpe performance measure over one- and five-year time horizons using quarterly returns. They found out that short-term performance persistence is present, but on the other hand limited to the high-yield bond sub-sample. Furthermore, performance persistence is no more likely among funds that had no changes in management than in funds that changed managers in the examined period. Finally, they concluded that in the case of high-yield bond funds, the risk-adjusted performance is inversely related to the portfolio performance.

Redman et al. (2000) in their research work examined the risk-adjusted returns using the Sharpe's index, the Treynor's index and the Jensen's alpha for five portfolios of mutual funds and for three time periods (1985 through 1994, 1985-1989, and 1990-1994). The benchmarks used for comparison were a proxy for the stock market and a portfolio composed of mutual funds that invest in U.S. issued stocks. The results indicated that there are potential diversification benefits to adding global funds to portfolios of domestic mutual funds.

Chamorro (2000) applied a different approach in mutual fund evaluation based on option pricing theory. They showed how to apply the portfolio insurance dynamics to the evaluation of 35 mutual funds over the period 1991-95. They estimated the insurance premia for Spanish funds of diverse composition and computed their net-of-downside-risk returns in order to rank the funds according to these criteria. Although the results concerning the ranking by risk were not too surprising, those related to the classification by covered return were more efficient.

Ahmed (2001) evaluated various models in order to determine their efficiency in forecasting correlation among 202 equity mutual funds over the period 1979 to 1999. Correlation among funds is a very important determinant of portfolio risk. Precisely, this study forecasts mutual fund correlation using eight models (historical, mean and index models). According to their results, a Multi-Style Index, a Dynamic model and the Fama-Frenceh 3-Factor model presented the lowest prediction errors. Furthermore, the relative ranks of Multi-Style Index and Fama-French 3-Factor models have lower dispersion across different forecasting time period and in sub-samples of funds belonging to similar or different style categories. The inference of these results is important for managing the performance of a portfolio of mutual funds.

Gruber (2001) in his study identified the risk structure of mutual fund returns for 270 funds over the period 1985-1994 (in factor analysis) and for 274 funds over the period 1985-1995 (in cluster analysis). Precisely, he used a four-index model employing the S&P Index, and publicly available size, growth and bond indexes in order to examine what influences generate mutual fund returns and develop a model for measuring performance. He used factor analysis and proved that a fifth index appears to be present. In the case where he tested a publicly available index of growth mutual fund performance he found out that it explains a large proportion of the residuals from a four-index model. Finally, the data suggested that cluster analysis could be best used as an added influence to the based model. On the other hand, adding an index based on the dividend yield value index to the base model with a Morningstar Growth Fund Index explains correlation in a better way.

Zopounidis and Pendaraki (2002) into their study presented an integrated multicriteria decision aid methodology for the portfolio selection and composition problem in the case of equity mutual funds over the period 1997-1999. The methodology used consists of two stages. In the first stage the mutual funds are ranked according to their performance through PROMETHEE II method (originated from the field of the outranking relations of MCDA) based on several different weighting scenarios, in order to construct a portfolio consisting of a limited set of the best funds. In the second stage of this methodology it was applied a continuous MCDA technique through a goal programming formulation in order to solve the mutual funds portfolio composition problem specifying the proportion of each fund in the constructed portfolio. The proposed integrated approach constitute a significant tool that can be used to provide answers to two vital questions: (a) which funds are the most suitable to invest, and (b) what portion of the available capital should be invested in each one of these funds.

4 Summary and Concluding Remarks

Although the topic of mutual funds evaluation has been ignored for many years, it recently received considerably attention. A brief overview of the main mutual fund performance measures is illustrated in this study. Moreover, the restrictions that are observed in the application of the performance measurements adjusted to risk are described. The empirical evidences of the studies that have been done regarding the mutual fund performance evaluation and other related areas of the mutual fund industry are also presented.

Some very interesting research topics, that are presented, refer to: (a) the analysis of the nature and definition of risk in the portfolio selection, (b) the investigation of persistence in funds' performance, (c) the predictability of mutual funds, (d) the existence or lack of superior performance of investment managers or even the correlation between selectivity and timing abilities, (e) the examination of performance measurements with or without benchmarks, etc.

The empirical literature upon the evaluation measurements of the performance of mutual fund portfolios referred to Treynor index (1965), Sharpe's index (1966), Jensen's performance index (1968), Treynor-Mazuy model (1966), Henriksson-Metron model (1981), the CAPM, and several optimization models, etc. Eventhough these performance measurements adjusted to risk have been widely used in the assessment of portfolio performance, researchers have noted several restrictions in their application, such as the use of a proxy variable of the theoretical market portfolio that can be criticized as inadequate, the evaluation of the performance of an investment manager for long and not short time periods, the acceptance of the assumption of borrowing and lending with the same interest rate, the validity of the CAPM, the consistency of the performance of investment managers over time, etc. There are alternative approaches that provide the requisite methodology framework in handling the problem of portfolio selection and management through a realistic and an integrated methodology generated through multicriteria decision aid, neural networks, etc.

References

Ahmed, P. (2001). "Forecasting correlation among equity mutual funds", *Journal of Banking and Finance*, vol. 25, 1187-1208.

Bello, Z. Y., and Janjigian, V. (1997). "A reexamination of the market-timing and security-selection performance of mutual funds", *Financial Analysts Journal*, vol. 53 (5), 24-30, (September/October).

Brockett, P.L., Charnes, A., and Cooper, W.W. (1992). "Chance constrained programming approach to empirical analyses of mutual fund investment strategies", *Decision Sciences*, vol. 23, 385-403.

Brown, S. J., and Goetzmann, W. N. (1995). "Performance persistence", *The Journal of Finance*, vol. L (2), 679-98.

Carhart, M. (1997). "On persistence in mutual fund performance", *The Journal of Finance*, vol. LII (1), 57-82, (March).

Carhart, M. (1995). "Survivor bias and mutual fund performance", *Working Paper*, School of Business Administration, University of Southern California, Los Angeles, Cal.

Chamorro, J. M., and Jose M. Pèrez de Villarreal. (2000). "Mutual fund evaluation: a portfolio insurance approach. A heuristic application in Spain", *Mathematics and Economics*, vol. 27, 83-104.

Chang, E. C., and Lewellen, W. G. (1984). "Market timing and mutual fund investment performance", *Journal of Business*, vol. 57 (1), 57-72.

Chiang, W.C., Urban, T.L., and Baldridge, G.W. (1996). "A neural network approach to mutual fund net asset value forecasting", *Omega, Int. J. Mgmt Sci.*,vol. 24 (2), 205-215.

Colson, G., Zeleny, M. (1979). "Uncertain prospects ranking and portfolio analysis under the condition of partial information" in Verlag Anton Hain [ed.]: *Mathematical Systems in Economics* 44, Maisenheim.

Cumby R.E.; Glen J.D. (1990). "Evaluating the performance of international mutual funds", *The Journal of Finance*, vol. XLV (2), 497-521.

Dahlquist, M., Engstrom, S., and Soderlind, P. (2000). "Performance and characteristics of Swedish mutual funds", *Journal of Financial and Quantitative Analysis*, vol. 35 (3), (September), 409-423.

Elton, E.J., Gruber, M.J., Das, S., and Hlavka, M. (1993). "Efficiency with costly information: A reinterpretation of evidence from managed portfolios", *The Review of Financial Studies*, vol. 6 (1), 1-22.

Elton, E.J., Gruber, M.J., and Blake, C.R., (1996). "The persistence of risk-adjusted mutual fund performance", *Journal of Business*, vol. 69(2), 133-157.

Fama E. F., and French, K.R., (1993). "Common risk factors in the returns on bonds and stocks", *Journal of Financial Economics*, vol. 33, 3-53.

Fama E. (1972). "Components of investment performance", *Journal of Finance*, vol. 27 (2), 551-567.

Fillipas, N., (1999). "*Mutual funds and stock exchange environment*", Ed. Globus Invest, Athens, (in Greek).

Friend, I., Brown, F., Herman, and E., Vickers, D. (1962). "A study of Mutual Funds", *U.S. Securities and Exchange Commission.*

Gallo, J. G., and Swanson, P. E., (1996). "Comparative measures of performance for U.S.-based international equity mutual funds", *Journal of Banking and Finance*, vol. 20, 1635-1650.

Goetzmann, W.N., and Ibbotson, R. (1994). "Do winners repeat? Patterns in mutual fund return behavior", *The Journal of Portfolio Management*, (Winter 1994), 9-18.

Grinblatt, M., and Titman, S. (1989). "Mutual fund performance: An analysis of quarterly portfolio holdings", *Journal of Business*, vol. 62(3), 393-416.

Grinblatt, M., and Titman, S. (1992). " The persistence of mutual fund performance", The Journal of Finance, vol. XLVII (5), 1977-84.

Grinblatt, M., and Titman, S. (1993). "Performance measurement without benchmarks: An examination of mutual fund returns", *Journal of Business*, vol. 66 (1), 47-68.

Grinblatt, M., and Titman, S., and Wermers, R. (1995). "Momentum investment strategies, portfolio performance, and herding: A study of mutual fund behaviour", *The American Economic Review*, vol. 85 (5), 1088-1105.

Gruber, M. J. (2001). "Identifying the risk structure of mutual fund returns", *European Financial Management*, vol. 7 (2), 147-159.

Hendricks, D., Patel, J., and Zeckhauser, R.(1993). "Hot hands in mutual funds: Short-Run Persistence of relative performance, 1974-1988", *The Journal of Finance*, vol. XLVIII (1), 93-130.

Henriksson, R.; Merton, R. (1981). "On market timing and investment performance", *Journal of Business*, vol. 54(4), 513-534.

Henriksson, R. (1984). "Market timing and mutual fund performance: An empirical investigation", *Journal of Business*, vol. 57 (1), 73-96.

Hurson, Ch., Zopounidis, C. (1995). "On the use of Multicriteria decision aid methods for portfolio selection", *Journal of Euro-Asian Management*, vol. 1/2, 69-94.

Indro, D.C., Jiang, C.X., Patuwo, B.E., and Zhang, G.P. (1999). "Predicting mutual fund performance using artificial neural networks", *Omega*, vol. 27, 373-380

Ippolito, R. A., (1989). "Efficiency with costly information: A study of mutual fund performance", *Quarterly Journal of Economics*, vol. 104, 1-23.

Jensen, C.M. (1968). "The Performance of Mutual Funds in the Period 1945-1964", *Journal of Finance*, vol. 23, 389-416.

Jensen, C.M. (1969). "Risk, the pricing of capital assets, and evaluations of investment portfolios", *Journal of Business*, vol. 42, 167-247.

Kahn, R. N., and Rudd, A. (1995). "Does historical performance predict future performance", *Financial Analysts Journal*, (November-December), 43-53.

Koh, F., Phoon, K., and Tan, C., (1993). "Market timing abilities of fund managers: Parametric and non-parametric tests", *Journal of Business Finance & Accounting*, vol. 20 (2) (January), 155-166.

Lehmann, B.N; Modest, D.M. (1987)."Mutual Fund Performance Evaluation: A Comparison of Benchmarks and Benchmark Comparisons", *The Journal of Finance*, vol. XLII (2), June, 233-265.

Litner, J., (1965). "The evaluation of risk assets and the selection of risky investments in stock portfolios and capital budgets", *Review of Economics and Statistics*, vol. 47, 13-37.

Malkiel, B. G., (1995). "Returns from investing in equity mutual funds 1971-1991" *Journal of Finance*, vol. 50, 549-572.

Markowita, H., (1952). "Portfolio selection", *Journal of Finance*, vol. 7 (March), 77-91.

Modigliani, F., and Modigliani L. (1997). "Risk-adjusted performance", *Journal of Portfolio Management*", vol. 23 (2) (Winter). 45-54.

Milonas, N. (1999). "*Greek mutual funds*", Ed. A. N. Sakoula, Athens, (in Greek).

Morey M.R., and Morey R.C. (1999). "Mutual fund performance appraisals: a multi-horizon perspective with endogenous benchmarking", *Omega, Int. J. Mgmt Sci.*, vol. 27, 241-258.

Murthi, B.P.S., Choi, Y.K., Desai, P. (1997). "Efficiency of mutual funds and portfolio performance measurement: A non-parametric approach", *European Journal of Operational Research,* vol. 98, 408-418.

Pendaraki, K., & Zopounidis, C. (2000). "Evaluation of equity mutual funds' performance using a multicriteria methodology", *Proceedings* of Decision Making Conference under Uncertainty in the Global Environment of the 21st Century, VII Congress of SIGEF, Greece (Chania, Crete), (September), 77-80.

Philpot, J., Hearth, D., and Rimbey, J. (2000). "Performance and management skill in nonconventional bond mutual funds", *Financial Services Review*, vol. 9, 247-258.

Redman, A. L., Gullett, N. S., and Manakyan, H. (2000). "The performance of global and international mutual funds", *Journal of Financial and Strategic Decisions*, vol. 13 (1), (Spring), 75-85.

Sharpe, W.F. (1966). "Mutual Fund Performance", *Journal of Business*, vol. 39, 119-138.

Simons, K. (1998). "Risk-adjusted performance of mutual funds", *New England Economic Review*, (September/October), 33-48.

Treynor , J.L. (1965), "How to rate management of investment funds", *Harvard Business Review*, vol. 43, 63-75.

Treynor, J.L., and Mazuy, K.K. (1966), "Can mutual funds outguess the market", *Harvard Business Review*, (July/August), 131-136.

Wermers, R. (2000). "Mutual fund performance: An empirical decomposition into stock-picking talent, style, transactions costs, and expenses", *The Journal of Finance*, vol. LV, (4), (August), 1655-1703.

Zopounidis, C., and Pendaraki, K. (2002). "An integrated approach on the evaluation of equity mutual funds' performance", *European Journal of Business and Economic Management*, (in press).

Risk Logistics in Banking Management

Risk Logistics in Banking Management

C.D. Tarantilis*, C.T. Kiranoudis

Section II: Process and Systems Analysis, Design and Development,
Department of Chemical Engineering, National Technical University of Athens,
GR 15780, Athens, Greece

Summary:
Up to the present day, routing of bank courier deliveries has been treated as a classical distribution problem, formulated as the well-known Capacitated Vehicle Routing Problem (CVRP). However, although the solution of the CVRP plays an important role in the efficiency of the operational planning of distribution management procedures of a bank, its classical formulation does not take into consideration the most important routing factor of a bank-related distribution operation: the safety of deliveries. Hence, this paper presents an adaptive memory-based metaheuristic algorithm for designing safe delivery routes for a fleet of homogeneous armoured vehicles, departing and terminating at the central office of a bank, which serves a set of delivery offices. The goal of this algorithm is to find the solution (set of delivery routes) that mitigates the risk of the armoured vehicles being robbed.

Keywords: Risk Logistics; Vehicle Routing Problem; Metaheuristics; Banking.

1. Introduction

The management of banks schedule deliveries that involve large amounts of money or precious objects from the central office of a bank to a set of delivery offices on a daily basis.

Up to the present day, routing of bank courier deliveries has been treated as a classical distribution problem, formulated as the well-known Capacitated Vehicle Routing Problem (CVRP)[1]. According to this famous logistics problem, given the distances (in terms of length or time) between delivery offices, the goal is to find routes for the identical armoured vehicles with a fixed carrying capacity in such a way that the total distance travelled by the armoured vehicles is minimised;

---

* Corresponding author. Tel.: +1 301772-3147; e-mail: tarantil@chemeng.ntua.gr

- the fixed capacity of an armoured vehicle cannot be exceeded;
- a delivery office has pre-specified demand that must be satisfied;
- the demand of a delivery office is satisfied by exactly one visit of a single armoured vehicle;
- an armoured vehicle must leave and return to the central office;
- the proposed vehicle routing plan has to be completed within a pre-specified time interval.

However, although the solution of the CVRP plays an important role in the efficiency of the operational planning of distribution management procedures of a bank, since it offers great savings in its distribution costs, its classical formulation does not take into consideration the most important routing factor of a bank-related distribution operation: *the safety of deliveries*. It is clear that the savings in distribution costs are insignificant compared with the economic damage that a robbery involving an armoured vehicle of a bank can cause.

There is consequently a great need for developing routing algorithms aiming at finding safe delivery routes with respect to possible robbery attacks to armoured vehicles. Such an algorithm is presented in this paper.

2. The risk space

Since the objective of this application is to find safe delivery routes, the proposed metaheuristic has to design routes mitigating the risk of armoured vehicles being robbed. Hence, routing decisions must be made in a new space, the so-called risk space, where risk of a point, ρ (x,y), (risk of a vehicle being robbed on a point) is proportional to its distance to the closest police department.

The risk space is a Riemannian manifold because the metric is different from point to point in this space. Moreover, the length of road in the risk space is different from the length of the same road in the real space (due to the transformation of real space because of the risk function). Hence, contours of different risk level are determined around the police departments. The yellow, green and red contours represent the low, medium and high-risk areas respectively (see Figure 1). These contours are defined by employing the Multi-Start Shortest Path (MSSP) algorithm[2] that computes, for each network point, the shortest path tree associated with the minimum distance from the closest police department. Considering a typical road network of approximately 10,000 roads, the MSSP algorithm solves the problem of finding the shortest paths from one to all other nodes of the network, in 20 ms on a Pentium II/400 MHz-128MB RAM computer.

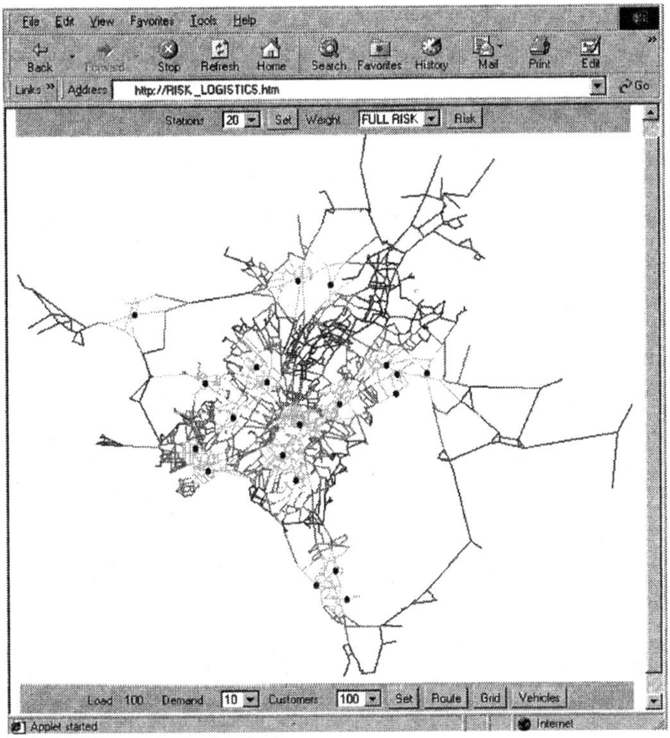

Figure 1. The low risk (yellow), the medium risk (green) and the high risk (red) areas for a vehicle being robbed, with respect to the distance of a point of the road network of Athens to the closest police department (black point).

According to the risk definition of this application, the closer to a police department a point that a potential robbery takes place is, the lower chances of taking place and being successful this robbery stands. This can be attributed to the following two reasons:

- Robbers' discouragement. The potential robbers will not consider to attacking within an area that is under heavy police surveillance;
- Rapid mobilization. Even in cases where a robbery takes place, police can mobilise its personnel rapidly and arrest robbers before they manage to get away.

Since the length of the road in the risk space is different from the length of the same road in the real space, we have to compute the length between the two nodes that define a projected road in the risk space (a road is considered as link which is defined by the up- and down-node respectively).

Because of the fact that risk of a point is not expressed by an analytical function in this application, but as a function of the distance between this point and the closest police department, the length of the projected road in the risk space is computed as follows: This solution procedure is based on the fact that there is always a point *C* on a road, as it is depicted in Figure 2(a), where the distance *l* between this point and the closest police department is equal to

$$l = l_1 + l_{AC} = l_2 + l_{AB} - l_{AC} \quad (1)$$

where l_1 and l_2 are the distances between the closest police department and nodes *A* and *B* respectively, l_{AC} is the distance between *A* and *C* and l_{AB} is the distance between *A* and *B*. The distances l_1, l_2 have already been computed by the MSSP algorithm and l_{AB} can be calculated because the coordinates of the nodes A and B are known.

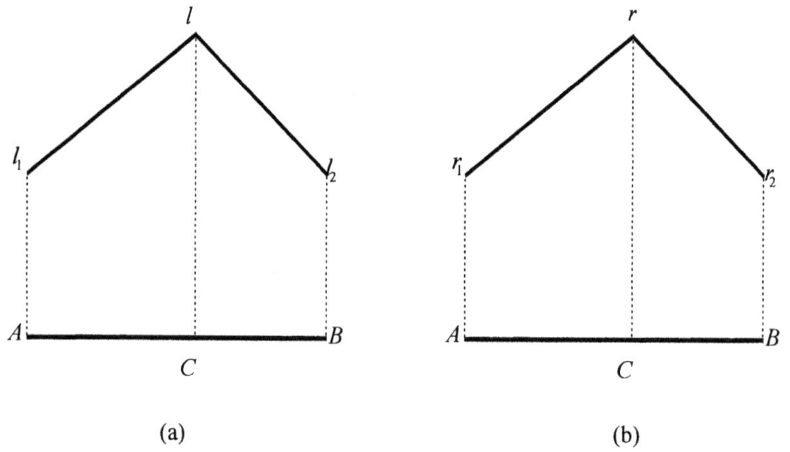

Figure 2. (a) The minimum distance required a vehicle reaching a network road point (b) The minimum risk of a vehicle being robbed

Due to the risk expression of this application, the point *C* is also the road risk equilibrium point where

$$\rho(l_1 + l_{AC}) = \rho(l_2 + l_{AB} - l_{AC}) \quad (2)$$

because the line segments located on both sides of the point can be reached by lower distance cost than the one necessary for this point.

After solving relation (2) with respect to l_{AC}, the distance (length) of the projected road in the risk space is computed as the sum of the distances (length) between points *A* and *C*, and points *B* and *C* respectively, since risk is linear with respect to distance from the closest police department, as it is shown by Figure 2(b).

In order to be able to employ the VRP algorithm, the Origin and Destination (OD) matrix must be constructed. The OD matrix consists of the shortest paths between delivery offices and between a delivery office and the central office of a bank. The computation of the shortest path tree associated with each and every one of the network's nodes requires the employment of the Dijkstra algorithm[3] that computes the shortest paths from a "source" node to all other nodes of the real road network. However, the repeated application of the Dijkstra algorithm finds the shortest path tree associated with each and every one of the network's nodes.

Provided that the shortest paths are known, the proposed metaheuristic is employed in order to find the solution (set of delivery routes) that mitigates the risk of the armoured vehicles being robbed.

3. The proposed metaheuristic algorithm

The routing problem of this application is examined on the road network represented as $R = (N, L)$, where $N = \{0,\ldots, n\}$ is a set of nodes (delivery offices), and L is the set of links. The central office of a bank is located at node 0. With every link is associated a non-negative distance matrix $C = (c_{ij})$, where c_{ij} is interpreted as a travel distance (length).

The CVRP is a problem of high computational complexity (*NP*-hard), so that in practical applications involving more than 75 customers, a large number of approaches focus on using heuristic approaches[4,5,6,7,8,9] rather than exact algorithms. The latter could give the global solution to small-impractical problems, however, the size of the vehicle routing problems, in terms of variables, grows with the square of customers (delivery offices) and proportionally to that to the number of vehicles, comprising a superstructure of possible available choices to incorporate in the routing schedule.

The proposed metaheuristic algorithm is a genetic solution procedure in the sense that it produces a new solution out of components of routes of previous good solutions. The components of routes used in this new algorithm *are sequences of nodes, called bones.*

The generation of a new solution (a set of routes) by the metaheuristic algorithm is based on the Adaptive Memory concept. Rochat and Taillard[10] introduced this concept to describe a pool of good solutions that is dynamically updated throughout the solution search process. Some components of these solutions are extracted from the pool periodically and combined to generate a new solution.

The main difference between the Rochat and Taillard method and our approach is that they attempt to intelligently combine routes instead of sequences of nodes (bones). More specifically, Rochat and Taillard use a small set of initial routes and create new solutions by using a probabilistic selection combined with a heuristic improvement procedure. In addition, routes are chosen from the pool by giving preference to routes belonging to good solutions and by ignoring routes that include delivery offices belonging to routes already extracted.

In contrust, the rationale behind the proposed metaheuristic algorithm is to extract bones (sequences of nodes) from the pool of good solutions, based on the criterion of "existence frequency" of these bones in the structure of the routes in the pool. We expect that the more frequently a bone is a part of those routes, the better chance of belonging to the best solution stands. Therefore, the extracted bones are used as a starting point to search for a higher quality solution.

Problem 7 - 75 customers

0	36	71	60	70	20	37	0		
0	4	45	29	5	47	48	68	0	
0	63	23	56	41	43	1	33	0	
0	32	50	18	55	25	9	0		
0	34	52	19	59	14	53	35	0	
0	2	28	61	69	21	74	30	0	
0	27	15	57	13	54	8	46	67	0
0	40	39	31	10	58	72	12	0	
0	7	11	66	65	38	26	0		
0	51	16	49	24	3	44	17	0	
0	6	73	42	64	22	62	75	0	

⟹

0	4	20	70	60	71	69	0		
0	45	29	5	37	36	47	48	75	0
0	51	63	23	56	41	43	33	0	
0	9	25	55	18	50	32	0		
0	27	15	57	13	54	52	34	67	0
0	30	74	21	61	28	2	68	0	
0	7	35	14	59	19	8	46	0	
0	12	72	39	31	10	58	26	0	
0	38	65	66	11	53	0			
0	16	49	24	3	44	40	17	0	
0	6	73	1	42	64	22	62	0	

Good solution = 913.23

Best solution =909.68

Figure 3. Tracing the common bones in routes between a good and the best solution

The reported best solutions for CVRP on the known set of vehicle routing problems[11] justify our idea since they are usually characterized by common sequences of nodes with the near best solutions on the same set of problems, as it is demonstrated in Figure 3.

The metaheuristic algorithm is described step by step as follows:

- PHASE I. Generation of the pool of good solutions or (Improvement phase)

In the first phase, the algorithm attempts to spread the search in order to examine over different regions of solution space *S* of the problem (diversification strategy) by generating *L* different solutions, used to produce the initial pool of routes. The heuristics, which are employed, are based on the weighted saving method proposed by Paessens[12] and the tabu search metaheuristic[13] respectively. In addition, in order that the algorithm excludes the lowest quality solution when the pool is updated, the R routes of the pool are sorted by increasing costs of the solutions where they belong.

The same methods are also used in order to improve the new feasible solution developed in the second phase of the method.

Regarding the tabu search method, it achieves a deeper exploration of the search space than the Paessens algorithm. Its main advantage is that it introduces the concept of tabu list T that keeps in memory not only the objective function value of the best solutions visited so far but also information on the itinerary through the last solutions visited.

So, after generating the solution s_1 by employing the Paessens algorithm, the tabu search performs as follows: the neighbourhood of s_1 is examined by using a blend of the 2-Opt , 1-1 and 1-0 Exchange[14] moves. These moves are distinguished in terms of exchanges performed to convert one tour into another, and they can be described as follows:

- 2-Opt move

Suppose a single route consists of the following set of nodes N_1 in the given order $N_1 = (0, 1, \ldots, k, 0)$, and let $A_1 = \{(i, i+1); (j, j+1)\}$ be a set of two edges in N_1 that form a criss-cross. 2-Opt move eliminates criss-cross and reverses a section of the route by deleting the branches $(i, i+1)$, $(j, j+1)$ and replacing them with its supplement (i, j), $(i+1, j+1)$ to reconstruct the route. In multiple routes, edges $(i, i+1)$, and $(j, j+1)$ belong to different routes but they form a criss-cross again. 2-Opt move is applied exactly in the same way as in the case of multiple routes. The move is granted when it is considered favourable for the performance of the entire algorithm in terms of objectives and constraints.

- 1-1 Exchange move

This move swaps two nodes from either the same or different routes. Consequently, if it is supposed that the initial tour consists of the set of nodes $(0, i\text{-}1, i, i+1, \ldots, j\text{-}1, j, j+1, \ldots, 0)$, the improved one is constructed as $(0, i\text{-}1, j, i+1, \ldots, j\text{-}1, i, j+1, \ldots, 0)$. The same procedure is conducted in the case of multiple routes. The move is granted when it is considered favourable for the performance of the entire algorithm in terms of objectives and constraints.

- 1-0 Exchange move

This move transfers a node from its position in one route to another position in either the same or a different route. Consequently, while the initial tour is $(0, \ldots, i, i+1, \ldots, j\text{-}2, j\text{-}1, j, j+1, \ldots, 0)$, the improved one is constructed as $(0, \ldots, i, j, i+1, \ldots, j\text{-}2, j\text{-}1, j+1, \ldots, 0)$. The move is granted when it is considered favourable for the performance of the entire algorithm in terms of objectives and constraints.

In considering the size of the tabu list T, the first t_l *moves* fill up the tabu list in ascending order (move giving the lowest cost in the first place).

The move m_{best} giving the best neighbor of s_1 is selected and the m_{best}^{-1} is examined whether belongs to the tabu list T.

- If the $m_{best}^{-1} \notin T$, then the m_{best} is accepted and the m_{best}^{-1} is inserted in T while the move giving the lowest cost is potted out of T;
- If the $m_{best}^{-1} \in T$, then the m_{best} is rejected and the reverse of the second best move is checked whether belongs to T;
- If $m_{best}^{-1} \in T$ and the m_{best} gives the best solution among the other moves whose the reverse of them belong to T, then the m_{best} is accepted and the m_{best}^{-1} is inserted in T.

The Tabu Search procedure is stopped when the number of iterations conducted is larger than the maximum number of iterations allowed.

The generation of new solutions by Paessens algorithm and Tabu Search is repeated by assigning random values from uniform distributions to f and g in the saving calculation

$$s(i,j) = d(0,i) + d(0,j) - g^{*}d(i,j) + f^{*}\left|d(0,i) - d(0,j)\right| \tag{3}$$

of the Paessens algorithm to generate L diversified solutions creating a set of R routes in the pool.

- PHASE II. Isolation of the promising bones and generation of the new solution

Based on the expectation that the best solution on a vehicle routing problem has usually common bones of routes with other near best solutions of the same problem, we try to compose a solution of highest quality, bone by bone, by isolating periodically the common bones from the dynamically updated pool throughout the solution search process and using them as base for re-starting and intesifying the search inside promising regions.

A partial solution is generated with respect to the following user control parameters:

- The number of delivery offices (nodes) composing the extracted bones
- The frequency that a bone must appear among the routes of the pool in order to be extracted from it.

It is obvious that the higher the frequency is, the more intensified the search procedure will be.

An example of a partial solution after the first iteration for a problem involving 200 delivery offices is demonstrated in Figure 4. The extracted bones are composed of bones consisting of three nodes. The area of the delivery offices in Figure 4 is proportional to the quantity ordered.

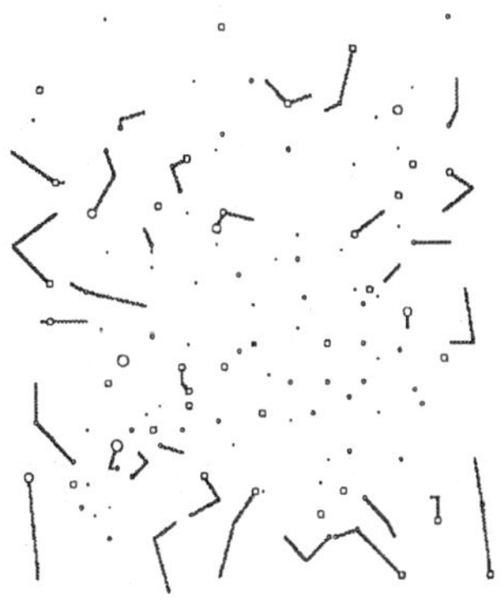

Figure 4. A partial solution generated after the first iteration by extracting three node-bones

Attention is paid so that the extracted bones do not involve the same customer. For this reason, if two or more different bones, which belong to different routes, satisfy the extraction criterion but they involve common delivery offices, the method chooses the bone for extraction from the pool that belongs to the route with the highest quality solution.

Regarding the delivery offices remaining in the pool subsequent to the extraction of a candidate bone, the proposed algorithm excludes from the routes only the delivery offices involved in the extracted bones by marking the part of each route in the pool that is common with the extracted bones.

Finally, a vehicle is assigned to each extracted bone from the pool and to each node that remains in the pool, in order to develop a new feasible VRP solution.

The proposed tabu search algorithm is summarized as follows:

- PHASE I

Step 1. Generate L initial solutions by employing the Paessens method and the standard version of tabu search successively.

Step 2. Insert R routes of the L solutions in a pool of solutions.

Step 3. Sort R routes by increasing costs of the solutions where they belong.

- PHASE II

Step 4. Extract the selected bones from the pool (based on user supplied control parameters, i.e. the number of delivery offices composing a bone, bone's frequency among routes in the pool).

Step 5. Exclude from the solution search process only the part of each route in the pool that is common with the extracted bones.

Step 6. Assign a vehicle to each extracted bone and to each node that remains in the pool in order to develop a new feasible solution

Step 7. Repeat from Step 1 to improve the new feasible solution by employing the Paessens and the tabu search algorithms respectively.

4. Case study

The proposed metaheuristic algorithm can be employed to solve real life problems, aiming at finding a set of delivery routes that mitigates the risk of the armoured vehicles being robbed.

Since the computational time of generating a vehicle routing plan (set of routes) is a factor of major importance for every real life distribution model, a standard parameter setting is used, which favoured small algorithmic computing times while at the same time exploited the diversification and intensification properties of the proposed algorithm, as they were described in the previous section. The parameters are set as follows:

- Tabu tenure equal to 20
- Pool size equal to 30
- Frequency of bones in the pool greater than 2
- Sequence of nodes composing a bone equal to 3
- The percentage of the 2-Opt move used to conduct local search is set equal to 35%
- The percentage of the 1-1 Exchange move used to conduct local search, is set equal to 35%
- The percentage of the 1-0 Exchange move used to conduct local search is set equal to 30%

The methodology described in the previous sections has already been applied once but since the management of the company demanded no exact routing and geographical data will be presented for proprietary and security reasons, we will use random locations of police departments, delivery offices and central office in order to implement our model in practice.

The problem takes places on the road network of Greater Athens area, as it is demonstrated in Figure 5, which involves more or less 10,000 roads. Twenty police departments, represented by black points, divide the entire network into regions of variable risk. One hundred delivery offices are represented by the blue points whose size is proportional to the required demand. The central office is represented by the red point. Each armoured vehicle starts at the central office and after satisfying the demand of each delivery office it visits, it returns back to the central office.

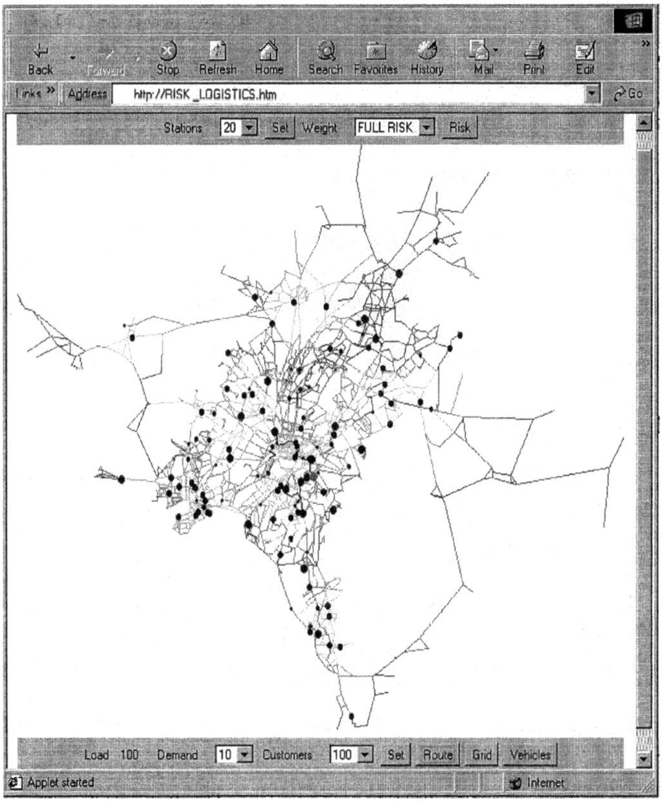

Figure 5. The examined case study

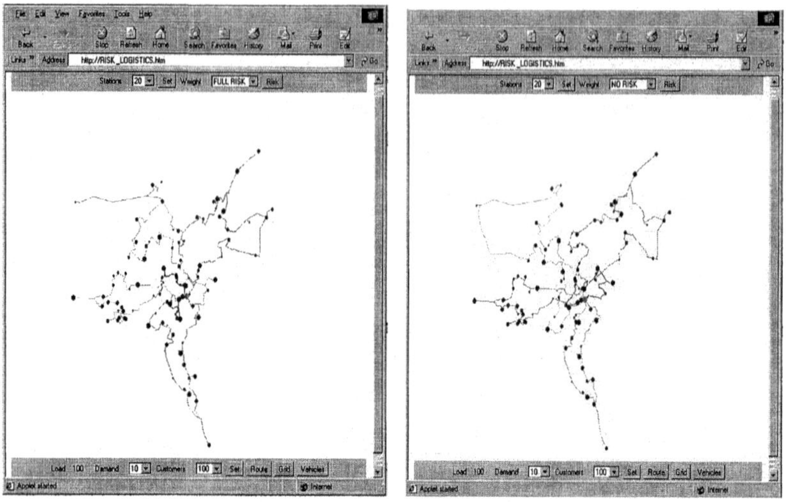

(a) (b)

Figure 6. (a) Mitigating risk (b) Minimising distance

The vehicle routing plan, which demonstrates the best sequence of deliveries with respect to minimisation of risk of vehicles being robbed (minimisation of total distance travelled by vehicles in the risk space) is presented in Figure 6(a). In contrust, the best sequence of deliveries with respect to minimisation of total distance travelled by trucks in the real (non-risk) space is demonstrated in Figure 6(b).

For this particular problem, the computational time of the proposed metaheuristic algorithm is equal to 6 seconds (Pentium II - 400 MHz). An instance of the problem was coded in Java 2.0 and run on Microsoft Internet Explorer 6.0. .

5. Conclusions

We have presented an adaptive memory-based metaheuristic algorithm for designing safe delivery routes for a fleet of homogeneous armoured vehicles, departing and terminating at the central office of a bank, which serves a set of delivery offices. The goal of this algorithm is to find a solution (set of delivery routes) that mitigates the risk of the armoured vehicles being robbed. The generation of new solutions by the proposed metaheuristic algorithm is based on the Adaptive Memory concept. According to this algorithm, a "pool" of good solutions is dynamically updated throughout the solution search process. Some *sequences of nodes* of these solutions are extracted from the pool periodically and combined to generate a new solution.

References

1. Laporte, G., Gendreau, M., Potvin, J-Y. and Semet, F. 2000. Classical and modern heuristics for the vehicle routing problem. International Transactions in Operational Research 7, 285-300.
2. Tarantilis, C.D., Kiranoudis, C.T., 2000. A GIS-based decision support system for risk logistics operations in banking management. Technical report CFD-12-00, Department of Chemical Engineering, National Technical University of Athens.
3. Winston, W.L., 1994. Operations Research. Applications and algorithms. International Thomson Publishing, California.
4. Salhi, S., Rand, G.K., 1987. Improvements to Vehicle Routeing Heuristics. Journal of the Operational Research Society 38(3), 293-295.
5. Osman, I.H., 1993. Metastrategy simulated annealing and tabu search algorithms for combinatorial optimisation problems, Annals of Operations Research 41, 421-451.
6. Taillard, E., 1993. Parallel iterative search methods for vehicle routing problems. Networks 23, 661-672.
7. Toth, P., Vigo, D., 1998. The granular tabu search (and its application to the vehicle routing problem), Working Paper, DEIS, University of Bologna, Bologna.
8. Tarantilis, C.D., Kiranoudis, C.T., 2001. Using a spatial decision support system for solving the vehicle routing problem. Information and Management (in press).
9. Tarantilis, C.D., Kiranoudis, C.T., 2001. A list based Threshold Accepting algorithm for the capacitated Vehicle Routing Problem. International Journal of Computer Mathematics (in press).
10. Rochat, Y., Taillard, E.D., 1995. Probabilistic diversification and intensification in local search for vehicle routing. Journal of Heuristics 1, 147-167.
11. Christofides, N., Mingozzi, A., Toth, P., 1979. The vehicle routing problem. In: Christofides, N., Mingozzi, A., Toth, P., Sandi, C. (Eds), Combinatorial Optimization, Wiley, 315-338.
12. Paessens, H., 1988. The savings algorithm for the vehicle routing problem. European Journal of the Operational Research 34, 336-344.
13. Glover, F., Taillard, E., Laguna, M., de Werra, D., 1993. Tabu Search. Annals of Operations Research 41, 4-32.

14. De Backer, B., Furnon, V., Shaw, P., Kilby P., Prosser, P., 2000. Solving Vehicle Routing Problems Using Constaing Programming and Metaheuristics. Journal of Heuristics 6(4), 501-523.

Investment Evaluation of a Suburban Coastal Transport System

E. Sambracos

Associate Professor of Transport Economics
Dept. of Maritime Studies
University of Piraeus
Greece

Summary:
This paper aims at determining the optimum viable solution of an investment on a suburban coastal shipping system in the area of Athens. More specifically it refers to the development of a sea transport system, alternative to the existing road one, that would connect Piraeus with the southern suburban coastal area of Athens. The best viable solution of such an undertaking is considered to be the one, which under the existing constraints maximises the total profit that derives from this investment. The variables used for the formation of the constraints are the number of vessels used, the routing and the price of the services. The article after presenting the methodology of the market analysis, focuses on the financing alternatives of the project and their impact on its economic efficiency and concludes with the best viable scenarios and optimum solution.

1 Introduction

The paper's main objective is to determine the optimum solution of an investment undertaking that includes an alternative transportation system, that is a suburban coastal transport system in the south region of Athens. Such a system is believed that can offer a possible solution to the transport problems of congestion, which characterise the existing overloaded road network of Athens.

Suburban coastal shipping has a special character. First there are not many examples of cities that use an analogous transport system. Secondly, the main competitor of sea transport is road transport or fixed track systems (train, tram). It should be mentioned that the current transportation system despite its deficiencies in terms of time, cost as well as externalities remains a strong travel alternative and consequently a constraint for the development of the proposed seaborne transport. This is due to the several advantages it offers to its users like door to door services, flexibility and speed. City of Athens disposes a road system as the main one satisfying the transport demand to and from the suburban areas.

The best economic solution for an investment of this kind is determined as the one that maximises the profit of the investor, under the constraints that the current transport system sets. In this context, in order to format the constrains of the system, a market analysis has been conducted taking all the above into consideration in order to examine the possible demand for such a system versus the existing road transport one (Greek Bank of Industrial Development, 1995). However, the question is whether an enterprise in this field would be economically viable and which would be the required demand of that service, in order to be able to equalise the marginal costs with the marginal revenue.

2 A methodology for the development of a suburban coastal transportation system in Athens urban area

2.1 The survey

It is obvious that the greatest part of the demand for sea transportation would come from the coastal suburbs that lie mostly in the eastern and southern part of Piraeus and their most important feature is that they constitute areas of permanent residence.

These suburbs include Alimos, Voula, Vouliagmeni, Glyfada, Elliniko, Kalithea, Moschato, Piraeus and P. Faliro. They constitute the districts of direct interest for this analysis (see figure 1). The case of the nearby districts that are situated northern than the coastal ones is also examined (Ag. Dimitrios, Argyroupolis, Vari, Ilioupolis, Koukaki, N. Smirni etc). This case obviously refers to the current car owners who can drive their cars up to the terminal station and then embark on the sea vessels after parking their cars (park and ride).

The first step in estimating the potential demand for a suburban coastal transportation system was to proceed to the determination of the total movements in the above mentioned area. This led to the development of the analogous Trip Distribution Model (Sambracos 2001) that shows the movements between the examined suburbs by private car and by bus. Data for the creation of the model was gathered by several origin – destination surveys that have taken place over the last years in addition with the records of the National Statistics Services on the population of the suburbs.

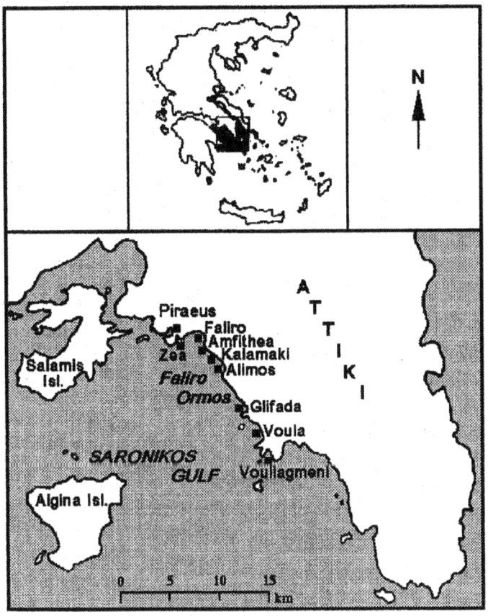

Figure 1: Suburban coastal shipping operation in the south region of Athens

The second step was to proceed to a Modal Split Model (Abacoumcin 1990, Giannopoulos 1981) and to determine the alternative means of transport that could satisfy the transport demand in and between the suburbs in question. In order to create such a model, the 'stated preference' method was used, a technique that uses and analyses the alternative scenarios, the questioned passengers state that they would prefer (Pearmain et.al, 1990, Bristow et.al, 2000). The questioned person is asked to choose between alternative situations/solutions, some of which may not yet be known to him, such as the proposed coastal transport system. The reason for using this method is mainly the fact that it can show results based on small samples (Koppelman et al, 1983). Thus the classic revealed preference method show disadvantages since it cannot be used in demand forecasting under existing conditions and cannot use qualitative criteria (such as the comfort of a transport means).

The survey was conducted with the use of questionnaires on June 1994 in the two main centres of interest, Piraeus and Glyfada. The questionnaire included 13 questions that had to do with:

- the used transport means,
- the origin and destination of the trip,
- the purpose of the trip,
- the frequency of the movement,

- the preference towards sea transport,
- the ability to use a car,
- the minimum and maximum time for the users of private cars and buses,
- the parking fees,
- the walking distance between the parking area or the bus station and the final destination.

Besides the questions there were six different scenarios, that referred to combinations of four main factors that determine the choice of a transport means. These are:

- the time of the trip (from door to door),
- the price of the ticket,
- the possible delays in the arrival of the vessel,
- the sea/weather condition.

Additionally, data regarding the sex, age and other demographic characteristics were taken into consideration.

In order to determine the demand allocation between the alternative transport means the Logit analysis was conducted (Ortuzar et. al., 1990) using as inputs the mentioned six scenarios and as variables the time of the trip (by bus/car and ship), the cost of the trip (by bus/car and ship), the delays (by ship), and the weather conditions of the coastal shipping alternative as well as the characteristics of the users (age, sex, purpose of the trip etc). Above them the time, cost of the trip and the weather conditions were statistically the most important variables.

2.2 Designing the system

In order to design the suburban coastal shipping system there are several factors of both quantitative and qualitative character to consider, using as data the outputs of the origin-destination model and the conducted survey (stated preferences).

The first factor was the ports that should be used to format the system, the number of stops between the main ports of Piraeus and Glyfada in combination with several levels of ticket prices. Three scenarios were examined:

1. Piraeus port – P. Faliro – Glyfada
2. Piraeus port – Zea port – P. Faliro – Glyfada and

3. Piraeus - Glyfada

The results from the survey showed that more mid stops would make the system unattractive and inefficient compared to the existing road one since they would increase the travel time. Additionally, more vessels would be required increasing therefore the cost of the system.

The price levels examined for each scenario referred to a sea fare of 300, 400, 500, 600 and 700 GRDrh respectively. The results of the stated preference model for a 12 hour and a peak hour transport work showed that the higher the number of mid stops the higher the demand for travelling. Based in that conclusion the third alternative was abandoned.

The second factor was the vessels type and the necessary number of vessels to cover the two and three mid stops alternatives. The number of vessels depends on the demand for each alternative on a 12h and peak hour (PD) basis, the time of the trip and consequently the maximum number of trips the vessel can perform (FB), the frequency of embarkations (FL), the capacity of the vessel (C = PD/FL) and equals to the ratio:

$$n = \frac{\mathrm{FL}}{\mathrm{FB}}$$

As for the type of the vessels, the choice was based on four factors:

- the purchase and operational cost
- the operational reliability, speed and flexibility
- the friendliness towards the environment
- the safety and comfort for the passenger

The proposed type of the vessel based on the above factors was the flying dolphin type. Its characteristics on speed (33knots/h) and capacity (110 pas.) prove that for the one mid stop scenario two vessels are requested and for the two mid stop scenario three vessels.

3 Defining the optimum solution

The best viable solution of such an undertaking is determined as the solution that maximises the Profit (P) of the undertaker. According to the economic theory maximising the Profit means maximising the difference between Revenues (R) and the Cost (C) that derive form the operation of the company. In other words it is:

$$\boldsymbol{P = R\text{-}C}$$

$$max\ (P) = max\ (R\text{-}C)$$

The constraints of the above profit maximisation consist of the existing alternative solutions, which derive from the market analysis on demand patterns towards the undertaking.

3.1 Cost of the suburban coastal transport system

The total cost, that such a transport company will face is a function of (Karvounis 2000:

- the investment cost (i),
- the operating cost (o)
- the financial cost (m)
- the depreciation cost (d)

Therefore we have,

$$C = f(i, o, m, d)$$

The total cost depends on the number of stops the vessels make. The more the stops the bigger the roundtrip time and consequently a bigger number of vessels is required in order to maintain a high level of service. This increases the investment cost since more capital is required to cover not only the acquisition of the vessels but also their operating expenses. Therefore, it is essential to determine the level up to which we must invest. There are different alternative scenarios that are based on the market analysis on origin – destination in order to determine the best alternatives for this system. The market survey already indicated two main alternatives using as a variable (i) the number of stops and the number of vessels to be employed:

- Scenario 1: one stop between the ports of origin and destination, using two vessels on pick hours
- Scenario 2: two stops between the ports of origin and destination, using three vessels on pick hours

That is ***Si***, where i= 1,2 stops

Considering these scenarios, the main cost elements are estimated as following:

a. Investment and operating costs

The estimation of the investment and operating costs is presented in the table 1 below. The investment cost consists of Founding & Organisation cost, Fixed investments (facilities etc), Working Capital, Unforseen expenses

The operating cost consists of Fuels, lubricans etc, Personnel, Office expenses, Maintenance Expenses, Port duties, Insurance, Advertisement

Table: 1 Investment and operating cost for the two scenarios (in 000GRDrh)*

Costs	Scenario 1	Scenario 2
a. Total investment costs	**1.150.000**	**1.500.000**
Founding - Organization	10.000	10.000
Fixed investments	940.000	1.240.000
Starting Capital	187.367	243.100
Unforeseen expenses	12.633	6.900
b. Total Annual Operating Cost	**562.100**	**729.300**
Fuels	281.000	337.000
Personnel	130.000	182.000
Office expenses	3.000	3.000
Maintenance Expenses	13.500	18.000
Port duties	45.000	80.000
Insurance	13.500	18.000
Advertisement	25.000	25.000
Unforeseen	51.100	66.300

*The estimation of the costs is based on market prices as of 1995

b. Depreciation cost

The cost of depreciation is estimated for both scenarios, using the depreciation rates, 4% for building and offices, 10% for the vessels and 20% for the rest equipment, as following:

- Scenario 1: annual depreciation for the first five years is 93,2mil.GRDrh and for the other five 91,2 mil.GRDrh.
- Scenario 2: annual depreciation for the first five years is 123,2mil.GRDrh and for the other five 121,2mil.GRDrh.

c. Financial cost

The financial cost for the development of this transport system deals with the financing of the investment costs for its development. Several scenarios of financing are proposed that include:

- different % of loaning and own capital

- several levels of loaning interest

3.2 Revenues of a suburban coastal transport system

The revenues from the operation of this system are a function of the demand (q) and the price of the ticket (p).

$$R = f(p, q)$$

The level of demand and the possible ticket price levels were examined during the market research. Although all price levels are possible the research showed that a range between 500GRDrh and 700GRDrh is the most viable one. Above this level the demand is low in favour of road transport and below it is economically unprofitable for the company. The two scenarios therefore are:

Sj, *where* j= 700GRDrh, 500GRDrh

It is assumed that for the working days the total revenues are:

$$2 \times N \times P$$

where, P is the price of the ticket and

N the corresponding total daily number of passengers.

For the weekends (2 X 52=104 days per year) the passenger traffic is estimated to be 70% of the traffic during working days. Therefore the total annual income is:

$$(261+0,7 \times 104) \times 2 \times N \times P = 667,6 \times N \times P$$

Using the above mentioned scenarios *Si* and *Sj*, four cases result:

-Case A: $S_{1,700}$ (1 mid-stop & price of the ticket 700GRDrh)

-Case B: $S_{2,700}$ (2 mid-stops & price of the ticket 700GRDrh)

-Case C: $S_{1,500}$ (1 mid-stop & price of the ticket 500GRDrh)

-Case D: $S_{2,500}$ (2 mid-stops & price of the ticket 500GRDrh)

Taking the above into consideration the total annual revenues are for each case:

- $S_{1,700}$: 913 mil. GRDrh.
- $S_{2,700}$: 961 mil. GRDrh.
- $S_{1,500}$: 828 mil. GRDrh.
- $S_{2,500}$: 875 mil. GRDrh.

In order to forecast the total revenues for the years to come there is the assumption that the traffic natural increase rate is 1,5% annually, therefore all the above amounts for the future year e should be multiplied by $1{,}015^{(e-1)}$ (e=1, the first year of evaluation).

4 Financial analysis of the system

4.1 Evaluation of the financial cost

The determination of the best financial solution that maximises the profit of the Coastal Transport Company is the next step of the investment evaluation of the system. The financial cost (m) of the undertaking is determined by two variables:

- **l**: the percentage (%) of a loan to cover the investment cost and
- **n**: the level of the interest rate

Therefore m is a function of l and n, or

$$m = f\,(l, n)$$

and so $$C = f\,(i, o, l, n, d)$$

All cost components have been determined for all S_{ij}. Assuming that the variables in the above function are l and n, we will try to maximise it by using all possible combinations of l and n.

In order to evaluate the financial efficiency **(Goulielmos 1997)** based on the above variables of this initiative we use the method of the Net Present Value and the Internal Rate of Return. Using a discount factor of 4%, the target is to find the alternative S_{ij} that gives the higher Net Present Value (NPV) as well as IRR.
Using the above variables we proceed to a 10 years simulation in order to format the equation that proves the relationship between the percentage of loan for all possible levels of loan interest and the NPV, IRR results. The results for all cases and the final equations are included for each case in table 2. The general form of the derived equations, as resulted from the simulation, is a first degree one, with the following formation:

$$Y = a - bX$$

Where Y = NPV

X = the percentage of loan at different levels of interest rate

a, b = constants

The main conclusion is that $S_{1,700}$ is the optimum solution, since it shows the greatest Net Present Value and Internal Rate of Return for all different levels of interest (Table 2). Regarding the financing of the project, it is concluded that in every level of the interest rate, the best solution is to self-finance the project.

Another important observation is that in $S_{1,700}$, the percentage of loan that makes the NPV equal or less than zero (which means that the project is economically a non viable solution) is the highest, in comparison with the other Cases. For example for $S_{1,700}$ (see table 2 and figure 2), in the case of a loan interest of 4% the company can take a loan that does not exceed 97,62% of the investment cost. In all other cases, this percentage is much lower, which means bigger financial risk and danger, in the case of inability to self-finance a significant proportion of the investment cost. The same conclusion can be derived for all levels of interest loan. Taking the above into consideration, the second viable solution is $S_{1,500}$, where for a 4% of loan interest, the total loan must not exceed 70,03% of the investment cost.

Table 2: Financial viability of the project for S_{ij}

	$S_{1,700}$		$S_{2,700}$	
Loaning Interest Rate	Equation (y=a-bX)	% of Loan that makes NPV=0	Equation (y=a-bX)	% of Loan that makes NPV=0
2%	Y=2.352.749-2.363.779X	99,53%	y=1.299.424-3.083.191X	42,14%
4%	Y=2.352.749-2.410.128X	97,62%	y=1.299.424-3.143.645X	41,33%
6%	Y=2.352.749-2.456.477X	95,78%	y=1.299.424-3.204.100X	40,55%
8%	Y=2.352.749-2.502.825X	94,00%	y=1.299.424-3.264.555X	39,80%
10%	y=2.352.749-2.549.174X	92,29%	y=1.299.424-3.325.009X	39,08%
15%	y=2.352.749-2.665.045X	88,28%	y=1.299.424-3.476.146X	37,38%
	$S_{1,500}$		$S_{2,500}$	
Loaning Interest Rate	Equation (y=a-bX)	% of Loan that makes NPV=0	Equation (y=a-bX)	% of Loan that makes NPV=0
2%	y=1.687.936 - 2.363.780X	71,41%	y=626.789 - 3.083.191X	20,33%
4%	y=1.687.936 - 2.410.128X	70,03%	y=626.789 - 3.143.645X	19,9%
6%	y=1.687.936 - 2.456.477X	68,71%	y=626.789 - 3.204.100X	19,56%
8%	y=1.687.936 - 2.502.825X	67,44%	y=626.789 - 3.264.555X	19,20%
10%	y=1.687.936 - 2.549.174X	66,21%	y=626.789 - 3.325.009X	18,85%
15%	y=1.687.936 - 2.665.045X	63,34%	y=626.789 - 3.476.146X	18,03%

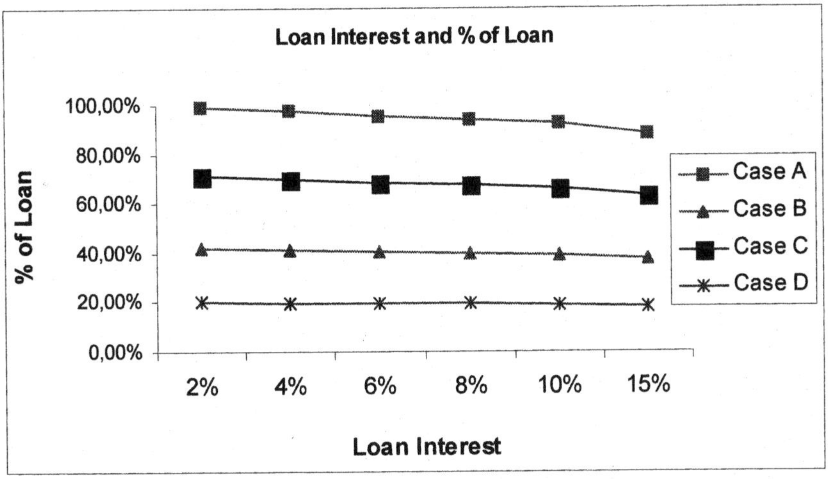

Figure 2: Financial viability of the project

4.2 Sensitivity analysis

Another important relationship is observed between the level of the loan and the economic effectiveness of the project (Theofanides, 1985). It is also essential to recognise the risk of such an investment (we choose the optimum solution $S_{1,700}$). Since the calculation of the costs is based on real data, while the approach of the revenues is based on demand forecasts, a sensitivity analysis is performed, on different levels of demand. The results are presented in figure 3.

Figure 3 shows that, the lower the loan, the lower the danger for all different levels of demand, which we normally expected. A 10% reduction in revenues has no impact on the viability of the project. On the other hand a loan of 30% or more of the investment cost shows great sensitivity if the revenues reduce for 20% and higher.

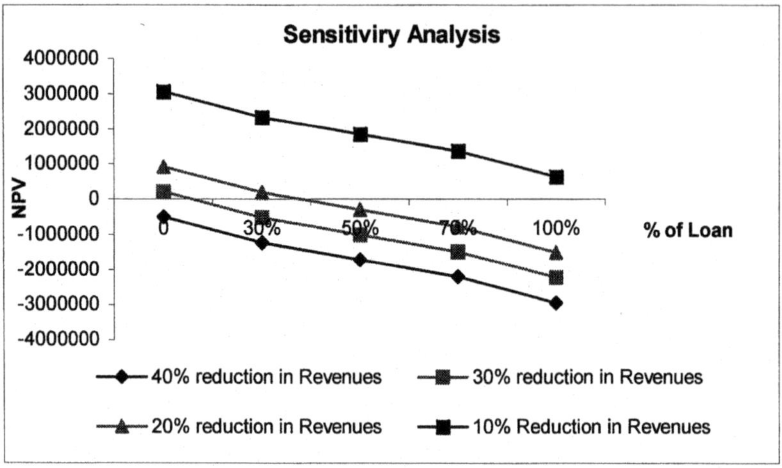

Figure 3: Sensitivity analysis for $S_{1,700}$

5 Conclusion

From the above analysis it is concluded that the implementation of a suburban coastal transport system in the southern Athens area is economically viable. This can be achieved with one or two mid stops and different price ticket levels. The optimum though economic solution is accomplished when we apply the smallest number of mid stops and the higher ticket price level. In the examined case study, an one mid stop system with a 700GRDRh ticket ($S_{1,700}$), is the most efficient financial solution since it shows the higher NPV and IRR. That means that the investor has bigger profit by offering the minimum service (1 mid. stop) with the highest price of the range examined (500 - 700 GRDrh). If he increased the stops and offer more service, and if he reduced the ticket price, then he would attract more customers. But, according to the above analysis the best solution is $S_{1,700}$, that means less service, less passengers and higher ticket price.

Thus, the percentage of Loan that makes NPV=0 is higher in case $S_{1,700}$, in every level of loan interest. That means that offering lower service with the higher price holds minimum financial risk for the investor. Additionally, the sensitivity analysis shows that the higher the level of uncertainty in the total revenues (reduction in demand) the lower the percentage of the loan, that the investor should take. From that point of view the case $S_{1,700}$, proves to be the most secure one.

References

Abacoumkin K. (1990) *Transport Systems Planning* , Symmetria, pp27-43

Bristow A., Nellthorp *J.* (2000) *Transport project appraisal in the European Union*, Transport Policy, Volume 7, Issue 1, pp51-60

Giannopoulos G. (1981) *Transport Planning and Traffic techniques* Athens pp.187-212

Goulielmos A. (1997) *Financing Maritime Companies* Publ. Stamoulis, Athens, pp. 358-365 (in Greek)

Greek Bank of Industrial Development (1994) *Investigation of the operation of a coastal shipping transport network between Piraeus and the South suburban area*"

Karvounis S. (2000) *Economic & Technical Studies*, Publ. Stamoulis (in Greek), pp 723-740

Koppelman F.S., Chu C. (1983) *Effect of Sample Size on Disaggregate Choice Model Estimation and Prediction*, Transportation Research Record 944, Washington D.C.

Ortuzar J, Willumsen L.G. (1990) *Modeling Transport*, John Wiley and Sons

Pearmain D., Kroes E., (1990) *Stated Preference Techniques: a guide to practice*, Steer Davies & Gleave Ltd, Hague consulting Group

Sambracos E. (2001) *Introduction on Transport Economics* Publ. Stamoulis, Athens, pp 248-251 (in Greek)

Theofanides S. (1985) *Manual for the Evaluation of Investment Projects*, Greek Bank of Industrial Development, Publ. Papazisis, pp. 256-260 (in Greek)

Author Index

Directory of Contributors

A. Anastasiou
University of Patras
Dept. of Management and Operational Research
Rio 26500 Patras, Greece

E. Barucci
Dipartimento di Statistica e Matematica Applicata all'Economia
Università di Pisa, Italy

A. Bouranta
University of Piraeus
80, Karaoli & Dimitriou
18534 Piraeus, Greece

A. Chevalier
ESCP-EAP
79, avenue de la République
75543 Paris Cedex 11, France

M. Daubie
Catholic University of Mons (Fucam),
Production and Operations Management Dept.
151 Chaussée de Binche
B-7000 Mons, Belgium

N. Dhillon
Management Development Institute
Sukhrali, Mehrauli Road, Sector 17
Guragaon 122 001 (Haryana), India

M. Doumpos
Technical University of Crete
Dept. of Production Engineering and Management
Financial Engineering Laboratory
University Campus
Chania 73100, Greece

M. Michalopoulos
Technical University of Crete
Dept. of Production Engineering and Management
Financial Engineering Laboratory
University Campus
Chania 73100, Greece

J. Mitropoulos
University of Patras
Department of Management and Operational Research
Rio 26500 Patras, Greece

K. Pendaraki
Technical University of Crete
Dept. of Production Engineering and Management
Financial Engineering Laboratory
University Campus
73100 Chania, Greece

E. Petridou
Aristotle's University of Thessaloniki,
Department of Economics,
Division of Business Administration, 54006
Thessaloniki, Greece

R. Renò
Scuola Normale Superiore
Pisa, Italy

E. Sambracos
Dept. of Maritime Studies
University of Piraeus
Piraeus, Athens, Greece

D. Singh
Management Development Institute
Sukhrali, Mehrauli Road, Sector 17
Guragaon 122 001 (Haryana), India

G. Dounias
University of the Aegean,
Dept. of Business Administration
Chios 82100, Greece

N. Glaveli
University of Macedonia, Dept. of Business and Administration,
54006 Thessaloniki, Greece

J. Gupta
ESCP-EAP
79, avenue de la République
75543 Paris Cedex 11, France

D. Hatas
Technical University of Crete
Dept. of Production Engineering and Management
Financial Engineering Laboratory
University Campus
Chania 73100, Greece

O. Ioannidis
National Bank of Greece
Thessaloniki, Greece

D.G. Karellas
Alpha Bank
Stadiou 40
105 52 Athens, Greece

C.T. Kiranoudis
Section II: Process and Systems Analysis, Design and Development,
Dept. of Chemical Engineering, National Technical University of Athens, GR 15780, Athens, Greece

K. Kosmidou
Technical University of Crete
Department of Production Engineering and Management
Financial Engineering Laboratory
University Campus
73100 Chania, Greece

A. Kouremenos
University of Piraeus
80, Karaoli & Dimitriou
18534 Piraeus, Greece

Y. Siskos
Technical University of Crete
Dept. of Production Engineering and Management
University Campus
73100 Chania, Greece

A. Sissouras
University of Patras
Department of Management and Operational Research
Rio 26500 Patras, Greece

Ch. Spathis
Aristotle's University of Thessaloniki
Dept. of Economics,
Division of Business Administration
54006 Thessaloniki, Greece

Ch. Staikouras
EFG Eurobank Ergasias
Strategy Investment and Planning Division
8 Othonos Street
10557, Syndagma Square
Athens, Greece

D.S. Stamoulis
Alpha Bank
Stadiou 40
105 52 Athens, Greece

O. Stathas
Economic and Business Management Dept.
Mediterranean Agronomic Institute of Chania
73100 Chania, Greece

K. Stavroulakis
Technical University of Crete
Dept. of Production Engineering and Management
Financial Engineering Laboratory
Chania 73100, Greece

C.D. Tarantilis
National Technical University of Athens
Department of Chemical Engineering
Section II: Process and Systems Analysis, Design and Development
GR 15780, Athens, Greece

E. Luciano
University of Turin
Dept. of Statistics and Applied Mathematics
Piazza Arbarello 8
10122 Turin, Italy

V. Manthou
University of Macedonia
Department of Applied Informatics
Thessaloniki, Greece

N. Meskens
Catholic University of Mons (Fucam),
Production and Operations Management
Department
151 Chaussée de Binche
B-7000 Mons, Belgium

M. Troege
ESCP-EAP
Département de Finance
97 av. de la République
75543 Paris, Cedex, France

M. Vlachopoulou
University of Macedonia
Department of Applied Informatics
Thessaloniki, Greece

G. Wood
City University Business School
Frobisher Crescent
Barbican Centre
London EC2Y 8HB, UK

C. Zopounidis
Technical University of Crete
Dept. of Production Engineering and
Management
Financial Engineering Laboratory
University Campus
Chania 73100, Greece